The Restoration of Ghirlandina Tower in Modena and the Assessment of Soil–Structure Interaction by Means of Dynamic Identification Techniques

T0256277

Built Heritage and Geotechnics
Series editor: Renato Lancellotta

Volume I

ISSN: 2640-026X
eISSN: 2640-0278

The Restoration of Ghirlandina Tower in Modena and the Assessment of Soil–Structure Interaction by Means of Dynamic Identification Techniques

Rossella Cadignani
Former Director of Historical and Monumental Building Division, Municipality of Modena, Italy

Renato Lancellotta
Former Professor of Geotechnical Engineering, Politecnico di Torino, Italy

Donato Sabia
Associate Professor of Structural Engineering, Politecnico di Torino, Italy

CRC Press
Taylor & Francis Group
Boca Raton London New York

CRC Press is an imprint of the
Taylor & Francis Group, an **informa** business

A BALKEMA BOOK

CRC Press/Balkema is an imprint of the Taylor & Francis Group, an informa business

© 2019 Taylor & Francis Group, London, UK

Typeset by Apex CoVantage, LLC

Library of Congress Cataloging-in-Publication Data
Applied for

Published by:
CRC Press/Balkema
Schipholweg 107C, 2316 XC Leiden, The Netherlands

First issued in paperback 2023

ISBN: 978-1-03-257073-0 (pbk)
ISBN: 978-0-367-18708-8 (hbk)
ISBN: 978-0-429-19778-9 (ebk)

DOI: https://doi.org/10.1201/9780429197789

Publisher's Note
The publisher has gone to great lengths to ensure the quality of this reprint but points out that some imperfections in the original copies may be apparent.

Contents

Preface   vi
About the authors   viii
Acknowledgments   ix

1   Introduction   1

2   Historical notes on the construction of the cathedral and
    Ghirlandina Tower   5

3   Previous restoration work: Ancient modifications and 1970s'
    interventions   35

4   Planning new investigation studies and restoration works   48

5   Soil as "material with memory": A key to explain settlements
    of the tower and the cathedral   78

6   A primer on soil–structure interaction and soil condition effects   95

7   Dynamic identification analyses: An approach to assess the real
    behavior of structures   109

References   131
*Appendix: Introductory notes on the leaning instability
of historic towers*   136
Index   143

Preface

This is the first of a series of volumes on Built Heritage and Geotechnics, intended to reach a wide audience: professionals and academics in the fields of civil engineering, architecture and cultural resources management and particularly those involved in art history, history of architecture, geotechnical engineering, structural engineering, archaeology, restoration and cultural heritage management and even the wider public.

Motivations rely on the fact that the preservation of built heritage is one of the most challenging problems facing modern civilization. It involves in inextricable patterns various cultural, humanistic, social, technical and economic aspects. The complexity of the topic is such that a shared framework of reference is still lacking among art historians, architects, structural and geotechnical engineers. This is proved by the fact that, although there are exemplary cases of an integral saving of any structural components with its static and architectural function, as a material witness of the knowledge, the culture and the construction techniques of the original historical period, there are still examples of uncritical confidence in modern technology which leads to the replacement of previous structures with new ones, which only preserve an iconic appearance of the original monument.

For these reasons publishing short books on specialized topics, like well-documented case studies of restoration work at one specific site or of a monument or detailed overviews of construction techniques, intended as a material witness of knowledge of the historical period in which the monuments were built, or specific conservation/preservation works, may be of great value.

The present volume, titled *The Restoration of Ghirlandina Tower in Modena and the Assessment of Soil–Structure Interaction by Means of Dynamic Identification Techniques*, provides within a single volume essential information on the history of the construction, the involved techniques, previous interventions and restoration work. It is proved how the interaction with the supporting soil may have had an important role because it explains the meaning of the corrections that masons did during construction, the pattern of settlements suffered by the tower and the cathedral and their interaction.

It is also discussed at length how the behavior of the tower during seismic events depends on the interaction with the supporting soil because such interaction influences both the seismic demand and the capacity of the system, and in this respect, this volume highlights the benefit derived by using identification techniques to capture and forecast this behavior.

In addition to the previously mentioned motivations, a single aspect should capture the interest of a wide readership: The cathedral and the Ghirlandina Tower were included in 1997 in the UNESCO World Heritage List, and it was recognized that

> [t]he creation process shared by Lanfranco and Wiligelmo is a masterpiece of human creativity, in which new dialectical relation between architecture and sculpture was

created in Romanesque art. The Modena complex bears exceptional witness to the cultural traditions of the XII century in northern Italy's urban society where its organization, religious character, beliefs, and values are all reflected in the history of the buildings.

The series editor
Renato Lancellotta

About the authors

Rossella Cadignani graduated in Architecture at the University of Florence (Italy). She worked in the municipality of Modena as Director of the Urban Development Planning area and later on of the Historical and Monumental Building Division. She assumed responsibility for the restoration work of Ghirlandina Tower and was a member of the Technical Committee of Modena UNESCO site. At present she is working as a freelance professional and is involved in historical research on the architecture of her city.

Renato Lancellotta was formerly Professor of Geotechnical Engineering at *Politecnico di Torino* (Italy) and is Chairman of the Committee on Preservation of Monuments and Historic Sites of the International Society of Soil Mechanics and Geotechnical Engineering. Professional experience refers to geotechnical studies for the preservation of built heritage, including the Leaning Tower of Pisa and the Shroud Chapel in Torino. He was a member of the Scientific Committee for the preservation of the Ghirlandina Tower and the Cathedral of Modena, the Santo Stefano complex of churches in Bologna, and the Giotto bell tower in Florence. He is also author of *Geotechnical Engineering* (Taylor & Francis, 2009).

Donato Sabia is Associate Professor of Structural Engineering at the *Politecnico di Torino* and is Head of the Materials and Structures Lab of the Department of Structural, Building and Geotechnical Engineering. His research is related to structural diagnosis, experimental modal analysis and dynamics identification of linear and nonlinear systems and seismic vulnerability of structures with special emphasis on masonry structures. He has been involved in dynamic identification studies of the Ghirlandina Tower and the Cathedral of Modena. He is a member of the Scientific Committee of the UNESCO site of Modena and of the Commission of the *Opera Primaziale di Pisa* to monitor the leaning behavior of Pisa Tower after the stabilizing interventions.

Acknowledgments

The authors are indebted to Dr Eng Renato Maria Cosentini for his assistance in preparing the ground response analyses and related figures in Chapter 7, and to Dr Veronica Padovani for providing the English version of Chapters 2, 3 and 4.

Chapter 1

Introduction

Historic masonry towers are an important part of the world cultural heritage, as they attest to the evolution of artistic expression and building techniques in a time span of several centuries. The *fresco* of Saint Theodore protecting Pavia is an impressive and remarkable picture of a skyline dominated by tall private towers (more than 40 built in the Middle Ages). This was also the case in Bologna, San Gimignano, Florence, Lucca, Pisa, Genova, Ascoli Piceno, Pistoia, Bergamo and many other Italian cities during the 10th and 11th centuries, when an impressive development of private towers took place as a tangible sign of the economic and social power acquired by individual citizens and merchants. The skyline of Florence was dominated by some 150 towers; in Bologna, private towers reached about 180 in number. In San Gimignano, about 60 km south of Florence, there is evidence that 72 towers were present in the 13th century, when the city lay along the *Via Francigena*; 14 of them survive and give us an idea of the original character of the urban ambience, so distinctive to be referred to as "the civilisation of the towers".

All these examples, and many others, tell us that historic masonry towers are part of our history and culture; they not only are a distinctive feature of the old town center, but quite often they are also the symbol of the town and represent the local community as a whole.

The preservation of historic towers requires a deep understanding of their structural response and the reasons that allowed them to survive over the centuries. It seems in fact evident that the towers we observe today survived the initial period in which they could have been not so far from a *bearing capacity collapse*, due to *lack of strength* of the soil. Delay or interruption of the building process allowed the foundation soil to improve its strength and the tower to be successfully finished. And due to uneven settlements, some of them appear today to survive at an alarming angle of inclination. This highlights the danger of a *leaning instability*, that increases if there is a *lack of stiffness* of the soil. In this respect, the Leaning Tower of Pisa represents a powerful example because the preservation of the tower was recognized as being a problem of leaning instability.

In addition, there are also problems of structural nature interacting with these geotechnical aspects. They depend on the masonry behavior as a unilateral material and deserve special attention as proved by the collapse of the Campanile in Venice in 1902 and the Civic Tower in Pavia in 1989 (Binda et al., 1992; Macchi, 1993). And obviously there has always been a focused interest in the potential seismic vulnerability of historic towers, further increased by the recent earthquake in the Emilia Romagna region in May 2012, because many historical masonry *campaniles* (bell towers) collapsed despite the moderate magnitude (M = 5.9) of the seismic event.

By considering all the mentioned examples, two strictly related aspects merge: it is evident how the preservation of historic towers deserves special attention, but at the same time, it must be stressed that there are important constraints when devising remedial measures and intervention techniques, due to the need of preserving the integrity of the monument.

It has to be recalled that, quite often, the integrity requirement is only interpreted as the requirement of preserving the shape and the appearance of the monument. In reality, it also implies historic integrity by considering the changes the monument experienced with time, as well as material integrity, that means construction techniques, materials and structural scheme. Therefore, preserving integrity requires a multidisciplinary approach as well as to develop the attitude not to rush in deciding the stabilizing measures until the behavior of the monument is properly understood (see Brandi, 1977; Calabresi and D'Agostino, 1997; Calabresi, 2013; Carbonara, 2012; Viggiani, 2017; and the volume edited by Lancellotta *et al.*, 2017).

Furthermore, when considering the vulnerability of built heritage, we need to carefully distinguish between the "intrinsic vulnerability" and what can be called the "induced vulnerability" (D'Agostino, 2017). The first one is linked to the conception of the monument, the used materials, the construction techniques and all changes or modification that took place during the long life span. All these aspects represent the historic integrity of the built heritage and as such are material witnesses of knowledge for the historical period in which the monuments were built.

The "induced vulnerability" is on the contrary linked to the reduction in time of the quality of the construction due to decay of properties of the materials, the masonry apparatus and the mortar, just to give some examples.

The "intrinsic vulnerability" does not matter for the built heritage we observe today because we lost constructions with intrinsic defects. The essence of restoration is therefore to remove any cause of added vulnerability to enhance the behavior of the construction.

It requires a multidisciplinary approach by taking into account all the mentioned aspects, the history of the construction, the involved techniques, previous interventions and, in this respect, the interaction with the supporting soil may have an important role because it can explain the meaning of the corrections that masons did during construction and it influences the vibration frequencies of the tower when subjected to seismic or dynamic loads.

This volume is part of a series of short books on specialized topics, like well-documented case studies of restoration works at one specific site or of a monument, or detailed overviews of construction techniques. It is devoted to the restoration of the Ghirlandina Tower in Modena and its interaction with the cathedral and the supporting soil and discusses the benefit derived by using identification techniques to capture and forecast its behavior.

The cathedral and the civic tower were included in 1997 in the UNESCO World Heritage List, and it was recognized that

> *[t]he creation process shared by Lanfranco and Wiligelmo is a masterpiece of human creativity, in which new dialectical relation between architecture and sculpture was created in Romanesque art. The Modena complex bears exceptional witness to the cultural traditions of the XII century in northern Italy's urban society where its organization, religious character, beliefs, and values are all reflected in the history of the buildings.*

Therefore, it is of paramount importance as a starting point to illustrate the history of its construction strictly linked to that of the cathedral, as is done in Chapter 2.

The Ghirlandina Tower recalls the Lombard bell tower model, with a square base of 11 m per side, which lightens upward through the progressive increase in the number of windows. The tower exterior has five horizontal partitions consisting of arched cornices with decorated shelves that identify six different floors; however, this design does not correspond to the internal organization, and the path of the stairs interferes with some of the openings of the third and fourth floors.

Above the square tower shaft, there is a smaller octagonal drum and a high spire crowned by a golden copper sphere topped with a cross, for a total height of 89.32 m.

The tower is characterized by an overall tilt out of plumb of 1.54 m in the southwest direction, apparent even to the less attentive spectator (Cavani, 1903). And during the first construction phases, the tower already underwent important disruptions that probably demanded long periods of interruption of the work and subsequent adjustments to build the upper portion vertically, just as happened in the more famous tower of Pisa.

The inclination corrections determined an arcuate pattern of the walls, following a line that is broken at least in four parts, as many as the construction phases.

The issue of leaning was the subject of concern and careful observation even in the past centuries. In 1898, a lead weight, hanging from the spire down to the basement, was installed to measure the inclination, and since 2003 there is an instrumental system that monitors the tower and the cathedral and is periodically upgraded.

Because of its recognized importance, restoration works took place in the 1970s and in more recent years between 2007 and 2011, as illustrated in Chapters 3 and 4.

Once all these historical and architectural aspects have been illustrated, Chapters 5 enters into the more specific aspect of soil–structure interaction. In particular, factual data on total and differential settlements are first presented. Then, a short digression on soil mechanics is also included in this chapter to discuss the soil as "material with memory" and how these concepts have contributed to the explanation of the differential settlements suffered by the cathedral and the tower, as well as their interaction.

Collapse events occurred in the past and recent earthquakes in Italy (May 2012) have once again put into evidence the need to assess the long-term behavior of the monument as well as its seismic vulnerability. To properly capture the behavior of a structure, particularly of a historic tower, during seismic events it is of paramount importance to take into account the interaction with the supporting soil and the effects related to soil conditions, that is, how soil properties may influence the input motion at the base of the tower.

The reader who is unfamiliar with concepts of dynamics may find it difficult to grasp these aspects, even in a rather simplified version, and bearing in mind these difficulties, we considered beneficial to have the Chapter 6 as a primer on basic aspects and to provide a lexicon that can help to appreciate the more specialized topic addressed in Chapter 7.

Nowadays it is apparently possible to analyze any complex interaction problem using advanced numerical tools, which are increasingly available. However, the use of these numerical tools is not easy; they are computationally intensive, they require a high level of expertise and they often lead to a large variability in the results, especially if different codes are used. Therefore, although simplified, basic analytical approaches, as those based on a single degree of freedom model still play a fundamental role in the understanding of the actual behavior of systems in engineering practice.

In this framework, Chapter 7 deals with dynamic identification techniques, and it is proved as these techniques represent a powerful means to detect the role of soil–structure interaction on the behavior of the tower and its beneficial effects.

We mentioned that due to uneven settlements, many historic towers appear today to survive at an alarming angle of inclination. This highlights the danger of a *leaning instability*, that increases if there is a *lack of stiffness* of the soil and creep effects.

Even if this problem was not of concern for the Ghirlandina Tower, considering its importance (the Leaning Tower of Pisa represents a powerful example because the preservation of the tower was recognized as being a problem of leaning instability), we provided an Appendix with some notes on this topic.

Chapter 2

Historical notes on the construction of the cathedral and Ghirlandina Tower

2.1 The context

Modena is located at the center of the fertile Po Valley, along the *Via Emilia*. Its territory is essentially defined by the Secchia and Panaro Rivers, which have changed their paths and flows during centuries.

Established as a Roman colony in 183 BC over previous settlements, the city owes its current framework to the Middle Ages, after numerous and devastating floods that destroyed the ancient monuments and progressively confined the Roman remains under approximately 5 m of sediments (Lugli, 2017).

The *Piazza Grande* is its geographical, religious, political and economic heart. The medieval city grew from this ideal center, which the city hall overlooks together with the cathedral and Ghirlandina Tower (Fig. 2.1).

The construction of the tower and the cathedral was initiated to replace the previous church, which safeguarded the relics of Saint Geminianus, Modena's patron saint[1] and which had lain in a state of great disrepair.

At the time of the investiture controversy between papacy and empire to appoint local bishops, Modena's bishop's seat was vacant after a time interval in which power had been firmly in the hands of Bishop Eriberto.[2] It was a period of growing civic initiative, which led after a few years to the formation of the free commune.

In this context, the people together with the local clergy, but independently from the ecclesiastical and imperial powers, decided to build a new great church, as was happening in other European cities.

The project was entrusted to the architect Lanfranco,[3] who conceived a mighty church with three naves and an annexed bell tower, an expression of a new figurative model that

---

1 Concerning Saint Geminianus (312/397 CE), bishop of the city, there is scarce information. According to tradition he died on January 31. He was much loved by the local Christian population, and to him several miracles were attributed, among which the healing of the daughter of Emperor Jovian of Constantinople, whose story is carved on the architrave of the *Porta dei Principi* on the southern side of the cathedral (around 1110). The cult of relics as a tool for spreading Christianity (for other saints of the period), was promoted by Sant'Ambrogio in the second half of the fourth century.

2 Bishop Eriberto was excommunicated in 1081 for having joined the imperial party of the antipope Guibert of Ravenna.

3 Lanfranco, known as the author of the project and director of the works of the cathedral of Modena, is referred to as *mirabilis artifex* and *mirificus aedificator* in the epigraph placed in the central apse of the cathedral, and he is cited three times in the manuscript O.II.11 titled *Historia Fundationis Cathedralis Mutinensis. Relatio de Innovatione Ecclesie Sancti Geminiani ac de Translatione Eius Beatissimi Corporis*, better known as "*Relatio*".

Figure 2.1 Aerial view of the cathedral and the tower

The *Relatio* is one of the most important manuscripts kept in the Archive of the Chapter of the Cathedral of Modena. In addition to reporting numerous privileges granted to the "house" of Saint Geminianus, copies of notarial deeds and contracts, in the first pages the document narrates the beginning of the construction of the

influenced the development of the Romanesque in the Po Valley and whose decorative apparatus accompanied the rebirth of sculpture in Europe.

2.2 The construction of the cathedral

On the 23 May 1099, the excavation work of the foundations of the church began, and the first stone was laid as early as 9 June. The building developed around the existing church starting from the apses, and on 30 April the remains of the saint were moved as soon as a section was sufficiently completed and autonomous. The dedication of the altar took place on 8 October 1106[4] in the presence of the pope, while the work on the rest of the building was still in progress. Matilde of Canossa[5] was present at the ceremony, and whilst she had not had an active role in the decision to erect the new church, she was responsible for the pope's presence, perhaps to sanction the regained submission to Rome of the Modenese church.

The cathedral was consecrated on the 12 July 1184[6] by Pope Lucius III, but the work continued for several years, with various modifications and additions to the construction.

The main modifications to the original plan were operated by the "*Maestri Campionesi*",[7] a group of stonecutters and stonemasons active in the construction site after Lanfranco's period and until the 14th century.

This was a great event, especially when considering that the citizens of a city, which, at the time, counted 12,000 inhabitants at the most, wanted it, and therefore, it is well representative of the medieval preoccupation with eternal salvation.

The building stands out as highly innovative for the period due to Lanfranco's architectural choices and the role of decoration entrusted primarily to the sculptor Wiligelmo.[8] The

---

Romanesque cathedral, the transfer of Modena patron saint's mortal remains from the old to the new cathedral and the consecration of the altar dedicated to him, which took place in 1106. The author of the text is likely to be the canon Aimone.

4 The transfer of the relics took place on 30 April 1106, and on that occasion Matilde of Canossa suggested not opening the sarcophagus and waiting for the pope's passage to the city, scheduled a few months later. On the 7 October the sepulcher was opened in the presence of Pope Pasquale II, who, the next day, dedicated the altar to Saint Geminianus and granted the remission of sins for the worshipers.

5 Countess Matilde of Canossa (1046–1115) ruled over a vast territory that extended from Lazio to Lake Garda. During the investiture controversy, she first played a mediating role; then she was the main supporter of the papacy and of the reform of the church. The main episode of the contrast between the church and the empire, the humiliation of Henry IV in front of Pope Gregory VII, took place in 1077 about 50 km from Modena in the castle of Canossa.

6 See the inscription by the *Porta dei Principi* on the southern side of the cathedral.

7 The Campionese masters were family groups of stonemasons from the area of Campione in the northern Como area, called *magistri lapidum*, who worked on the construction of the Duomo and the tower, as is documented in the contracts stipulated from 1190 to 1208 between the *massaro* Aygi and Anselmo from Campione, with which he commits himself and his successors to work *in perpetuum* in the worksite, and with the 1244 contract between the *massaro* Ubaldino and Enrico da Campione (nephew of Anselmo) at least until 1319, when the Ghirlandina Tower was completed.

8 Wiligelmus (or Vuiligelme, as argued by Salvatore Settis in his essay "*W pro V. La lettera rubata*", 1985) is the author of many of the façade sculptures, among which the exquisite stories from the Genesis. He is mentioned in the inscription on the main façade, which recalls the laying of the first stone, whose text reads: "The construction of the house of the great Geminianus began when the constellation of Cancer took its course, while that of Gemini said farewell, five days before the idis of the month of June in the year of the Incarnation of God a thousand one hundred minus one. Now, by the work of your sculpture, it is clear, oh Wiligelmus, of how much honor you are worthy among the sculptors". The last part of the text was added later on the same slab.

strong relationship between architecture and sculpture is a distinctive characteristic of the new Romanesque art and the two artists, Lanfranco and Wiligelmo, created a true master-piece, regardless of the later modifications.

The church is approximately 25 m by 60.5 m and is equivalent to a rectangle of 96 Modenese ft by 232 Modenese ft[9] measured from the center of the apses. Ettore Casari (1984) noted that there is a correspondence between the measurements of the different parts of the church and the $I\sqrt{2}$ module, a refined measurement system that suggests Lanfranco's erudite knowledge.

From a geometrical standpoint, however, the structure presents several irregularities.

The façade (Fig. 2.2) shows a difference in width in the partitions of the arches corresponding to *Porta Regia* and *Porta della Pescheria* and a 28-cm difference in the horizontal alignments that is clearly visible on the front elevation. In the interior, the distribution of the pillars and columns in the central nave also presents irregularities. These anomalies may have been determined by the presence of obstacles or by the need for corrections of a different nature, such as differential settlements. It is likely that the previous church was still functioning during the construction of the new one and that it constituted quite an obstacle to the advancement of the work (Fig. 2.3). The builders also had to deal with the remains of a third church and, ultimately, with the construction of the bell tower that caused settlement and tilting of the apses, as discussed in Chapter 5. On this topic, scholars have been debating widely and with quite different deductions, to which reference is made (see Peroni, 1984; Serchia, 1984; Silvestri, 2017).

The church is liturgically oriented with the main front to the west on a small parvis, the southern side facing the large square and the Ghirlandina Tower to the north, near the apses.

The structure is made of brick masonry, with a stone external cladding. The materials were largely salvaged from the remains of ancient buildings of the Roman city, still visible or emerged after excavating in different locations around the urban area.

The internal subdivision is declared from the outside, on a façade that is divided into three parts by two tall pilasters that become narrower at half height and mark the naves (see Fig. 2.2). The roof is gabled with window openings on the clerestory. The transverse arches that form the aisles prosecute beyond the roof, and their sides are crowned with metopes decorated with acrobatic and fantastic creatures.[10]

The whole building is surrounded by a series of arcades on half-columns with magnificent capitals, each one includes a three-lancet window, which contributes to form a "*chiaroscuro*" on the walls that accentuates the volumes. Below there is a cornice with small shelved arches decorated with human or animal protome, similar to those of the tower.

The main façade has a rich decorative apparatus, work of Wiligelmo, that includes the elegant volumes of the four panels that narrate the stories of Genesis, as well as the complicated interweaving on the jambs of the portal that narrate the struggle of man to reach salvation (Fig. 2.4).

On this front there are three door openings. The central one is protected by a two-story porch supported by reused Roman lions, which seems designed specifically to mitigate the

---

9 Modenese measures of the period: perch = 3.38289 m (corresponding to 12 ft), arm = 0.5230482 m; foot = 0.2615241 m.

10 The original metopes are kept in the Lapidary Museum of the cathedral and were replaced with copies during the 1950s.

Figure 2.2 The façade of the cathedral

difference in height of the frames in the center of the façade but which is also an element of novelty, later adopted as a model in many other churches and not only in the Po Valley.

During the so-called Campionese period, several modifications were made to the façade, the most relevant of which being the opening of the large-rayed rose window with 24 rays, which asked for the reduction in height of the porch. From Palazzi (1988) surveys, it was

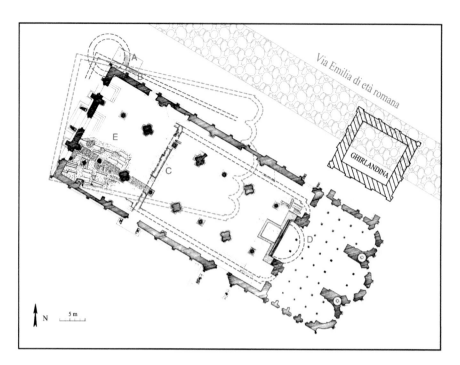

Figure 2.3 Reconstructive hypothesis (Labate, 2009) about the construction phases of the Cathedral of Modena. In brown is the theoretical reconstruction of the late medieval cathedral's perimeter with the circular baptistery; in light green, theoretical reconstruction of the pre-Lanfranco cathedral's perimeter based on the 1913 excavations; in dark green, new orientation of the pre-Lanfranco cathedral based on 1989 excavations; A – curved wall discovered in 1989 excavation and associated with the late medieval baptistery; B – straight wall found in 1989 excavation and associated with the northern perimeter of the pre-Lanfranco cathedral; C – foundation wall discovered in 1913; D – curved foundation wall discovered in 1919; E – 1913 excavation.

Figure 2.4 Wiligelmo, stories of Genesis

possible to verify that this insertion did not require the demolition of the whole façade tympanum, only of part of it, because the inclined brick course of the preexisting pitch is still intact. During the same intervention, the sculptures depicting the four evangelists were moved above the rose window, and an almond-shaped frame and a depiction of the Redeemer inside an almond-shaped frame was placed at the center.

The opening of the side doors determined the need to move the Genesis panels, which are no longer aligned; two are higher up, just above the new doors.

Turrets were placed at the sides of the front tympanum turrets and, in the center, the statue of an angel. The same model was then repeated on the back tympanum.

The southern façade (Fig. 2.5) overlooks the *Piazza Grande*, once called *Maggiore*, from which one can seize all the volumes and the relationship with the tall Ghirlandina tower.

Two doors open to the south; the first is the *Porta dei Principi*, decorated in the archivolt with scenes of Saint Geminianus's miracle, by Wiligelmo and his school. In 1944 the upper

Figure 2.5 The southern side with the *Porta Regia*

part of the porch was hit by a bomb during an air strike, destroying the *fresco* without affecting the precious reliefs.

Also, this side was affected by changes operated by the *Maestri Campionesi* during the 13th century: the *Porta Regia*, the superelevation of the presbytery and the false transept.

The *Porta Regia* is a large splayed door decorated with geometric and vegetal motifs, made of Rosso Ammonitico limestone from Verona, that has a characteristic rose tint. Two column-bearing lions support the loggia surmounted by the sculpture of a lion (a Roman piece), and the inside is supported by elegant columns decorated with Solomon's knots.

The presbytery is almost 2 m taller than the rest of the roof, and in the same section, on the side aisles, there is the tympanum of a "false" transept (1165–1184).

Beyond the *Porta Regia*, towards the apses, there is a pulpit with the symbols of the evangelists, a 16th-century addition, and a little further there is a 1442 bas-relief by Agostino di Duccio, a pupil of Donatello, with episodes from the life of Saint Geminianus.

Around the apses a trench highlights the settlements that took place during and after construction (see Chapter 5).

Above the single window of the central apse there is a plaque that celebrates the greatness of the architect Lanfranco and the foundation of the cathedral. The ancient Modenese measures are carved on the sides: the tile, the brick, the arm and the perch.

Just beyond the apses, a construction that no longer exists connected the cathedral to the Ghirlandina Tower, perhaps a sacristy that until the second half of the 16th century was the only entrance to the upper part of the tower.

This connection underwent several changes, some still visible on the façade and on the south side of the tower, where there are traces of an opening on the first floor, as well as the roof's impost. This first construction was demolished in 1338, replaced with two large arches, and then raised to contain a new room, which was demolished in 1902 when the current arches were built (Fig. 2.6).

Figure 2.6 The arches connecting the tower and the cathedral

The realization of the arches and the troubled history of the connection could have been determined not only by functional aspects[11] but also by the need to improve the stability of the two monuments.

Next to the second arch is the *Porta della Pescheria* (so called because it faced the bishop's fish market), whose jambs and archivolt, smaller than the architrave, are richly decorated with scenes from the Arthurian cycle.

The sacristy was built in 1470 attached to the northern side of the church, where the *Via Lanfranco* is today, and was connected with the house of canons that reached the *Via Emilia*.

When entering the church, the dominant white of the external stone cladding disappears, replaced by the warm color of the bricks (Fig. 2.7). The space is divided into three naves, with the central one taller, in whose walls are eight arches supported alternately by large poly-lobed pillars and stone columns with Corinthian capitals. From the pillars arise four acute arches transversal to the naves that intersect the walls and form as many spans in the side aisles. False matroneums lighten the central walls, while at the upper level, in the clerestory, there are two single windows for each aisle.

The cross vaults with false ribs that underline their shape were added in the mid-15th century. In the under-roof, there are traces of white and red decorations that were in place before the insertion of the vaults.

The church has no transept and the presbytery is raised above the naves. It is accessible from two side flights of stairs and ends with a magnificent marble parapet,[12] sculpted in the second half of the 12th century with scenes of the passion of Christ. The ambo is instead a

Figure 2.7 The interior of the cathedral

11 There was a need to separate the entrances: the canons and the tower keepers needed to access the upper part of the tower, while the employees of the archives of the municipality only the lower floors.

12 For a long time, it was believed that the parapet was the work of the Campionesi Masters; however, the chronology of the work, built between 1165 and 1184, has recently led scholars to consider it rather the work of a group of Paduan masters close to the sculptors who worked in the Parma baptistery.

Campionese addition from the beginning of the 13th century. The current configuration is the result of the restorations of the early 20th century, while it is probable that the original presbytery had an access staircase in a central position.

Below the parapet and just below the basilica floor level there is the crypt, which is supported by 28 columns with decorated capitals all different from each other, with shapes that can be referred to stone carvers that came before (or at least were different to) those of Wiligelmo's school. At the center of the main apse there is Saint Geminianus's tomb surrounded by an 18th-century stone balustrade.

Along the side aisles are numerous works of art, mainly from the 15th and 16th centuries, among which the most relevant are, on the left in the second bay, the *"Altare delle Statuine"* ("The Figurines Altar"), a grandiose 1440 terracotta polyptych by Michele da Firenze; in the back, before the stairs leading to the sacristy, there is the *Pala di San Sebastiano*, painted by Dosso Dossi in between 1518 and 1521; in the first bay on the right is the Bellincini chapel, built by Cristoforo da Lendinara between 1472 and 1476, as well as a large arch with "terracotta" architecture and *frescoed* vaults; and a little further is the 1527 elegant "terracotta" nativity scene by Antonio Begarelli.

In the following centuries the cathedral underwent many transformations following the changing taste in architecture, but none compromised the formal integrity of the building because the interventions were performed mainly on the ornaments rather than on the structure.

The most relevant one was the modification of the presbytery and of the accesses wanted in 1593 by the bishop, Canano, who had the panels of the pulpit separated to extend and adapt it to the needs of the liturgy. On the front, towards the naves, a wrought-iron balustrade was added.

Between the 18th and 19th centuries, a polychrome *"scagliola"* (a gypsum-based plaster) decoration that imitated marble in the crypt was realized, and an altar was erected on the Saint's tomb. Outside, between the *Porta Regia* and the *Porta dei Principi*, the painting of the *Madonna delle Ortolane* (*Our Lady of Greengrocers*) was covered with a marble altar in 1770, and on the same side, towards the churchyard, four shop booths were also built.

In the mid-19th century, a new culture began to spread regarding the conservation of ancient monuments and the revaluation of the "Romanesque" style. A debate about how churches should look began in Modena in 1877, and this led to the project of "restoration"[13] of the forms that the cathedral was supposed to have had in the 13th century, together with the removal of everything that had been modified over time and appeared in contrast with that idea. These are the years that marked the beginning of many interventions that concerned the base outside, the reduction of the connection with the house of canons, the demolition of the shops that were attached to the southern side and the removal of the Baroque altars.

In 1901, the isolation of the archiepiscopal palace that was adherent to the southern side towards the façade also began, as well as the demolition of the 1470 house of canons (Fig. 2.8) with the reopening of the street (*Via Lanfranco*) and the continuation of the interventions inside the cathedral with the final reconstruction of the parapet in today's form[14] in 1919.

---

13 A precursor of this approach was don Pietro Cavedoni in Modena, who wrote a booklet in favor of the reinstatement of the decoration, published in 1857.

14 The parapet was reassembled according to the interpretations of the scholar Tommaso Sandonnini (1849–1926), who supervised the works and wrote a scrupulous chronicle of the restoration of the Modena cathedral (1897–1925).

Figure 2.8 The demolition of the sacristy of the Duomo
Source: photo *Raccolte fotografiche modenesi* by Giuseppe Panini.

During this period a series of excavations were carried out inside the church, which uncovered the structures of the previous building.

Despite those valuable investigations and the more recent ones, there are still some aspects that are not completely clear regarding the real progress of the work and the changes that occurred during the 18th-century phases. The cathedral, as we see it today, is the result of an "interpretation", which even if debatable in the light of today, has nevertheless returned a monument in which the original work of Lanfranco and Wiligelmo still appears legible.

In 1997 the cathedral, the Civic Tower and the *Piazza Grande* were included in the UNESCO World Heritage List.[15]

2.3 The construction of the Ghirlandina tower

The tower, later named "Ghirlandina" probably because its balconies resembled garlands (see Baracchi and Giovannini, 1988), soon became the symbol of the city: what travelers saw from afar, standing out in the sky above all other buildings, clad in white (Fig. 2.1).

---

15 The 21st session of the UNESCO International Committee, held in Naples between 1 and 6 December 1997, recognized the monumental complex of Modena as a World Heritage Site.

The debate about the chronology of the Ghirlandina is still open because, unlike the cathedral, there are no ancient documents reporting the early construction phases. It is commonly accepted that the tower and the cathedral belong to the same construction plan and since the revenue managed by the *massaro* (manager) of Saint Geminianus concerned the factory of the church and the tower jointly and that the commune and the *Massaria*[16] had an equal role in the collection and the strategic management of resources,[17] it is reasonable to suppose that the building site was also unitary or strictly connected, if only for the logistical conditions in which the two buildings are.

On the other hand, we cannot exclude that the building of the tower, after the initial setting, may have been postponed for a few years, compared to the urgency of completing the portion of the cathedral that was used to translate the body of the saint or to wait for the settlement of foundations.

If the starting date is not certain, the completion date is: it was the 18 September 1319 by the hand of Enrico da Campione.

There are a few other dates that can give us a picture of how the works continued:

1169 is the date inscripted in the lower part of a decorated re-employed slab of Vicenza stone, placed on the east side under the second string course moulding, which was interpreted as indicating the beginning of the works on the third floor (see Lomartire, 2015; Serchia, 1984);

1217 when lightning struck one of the turrets placed at the top, so we can imagine a tower long completed with turrets placed in the corners of the last floor (the 5th) from which the guards could overlook the city walls;

1261 when the structure was raised from the 5th floor and the covering materials were purchased in Verona.

A useful datum to better understand the long construction process came from the photogrammetric survey performed in 2004[18] and from the laser scanning survey in 2008,[19] with which the geometry and the inclination were measured and a three-dimensional model was produced. There has been an almost homogeneous inclination of the walls toward the west, while towards the south there is a markedly curvilinear trend (see Fig. 2.9), typical of constructions that undergo leaning during the construction phase (Alfieri *et al.*, 2009).

An overall out of lead of 1.54 m was measured in the southwest direction, apparent even to the less attentive spectator.

Already, during the first construction phases, the tower underwent important disruptions that probably demanded long periods of interruption of the works and subsequent adjustments

---

16 The *fabbriceria*, in particular, since 1327, is an independent citizen institution; the *massaro* is a layman and acts in first person with rights and privileges, with the task of administering the funds donated and managing the construction of the Duomo and the tower. From the mid-16th century this figure lost its secular character, and the *massaro* was substituted by two canons who managed the administration of the Duomo, while the tower returned to the municipality that also took care of the restorations (see Baracchi and Giovannini, 1988).

17 See the extract from the statutes of the municipality of Modena, contained in the code of the chapter O.II.11, quoted in Saverio Lomartire (2015).

18 The survey was conducted by the University of Parma's Architecture Department.

19 The laser scanning survey was performed by the University of Modena and Reggio Emilia Department of Mechanical Engineering.

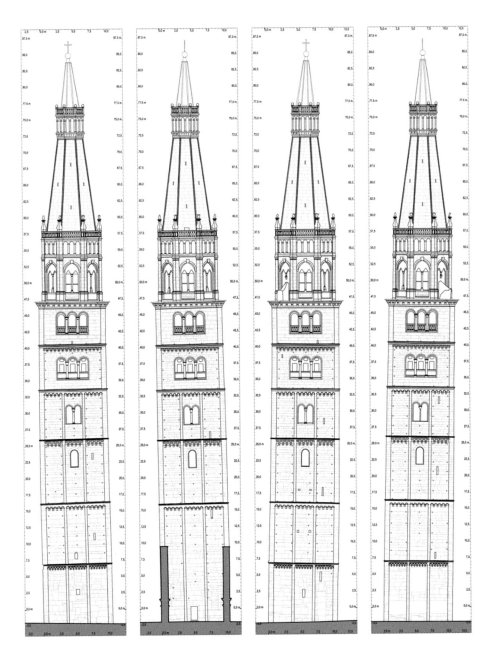

Figure 2.9 Front views. From the left: east view, south view, west view, north view

Source: Giandebiaggi *et al* (2009).

to build the upper portion vertically, just as it happened in the more famous tower of Pisa. The inclination corrections determined an arcuate pattern of the walls, following a line that is broken at least into four parts, as many as the construction phases.

The issue of leaning was the subject of concern and careful observation even in past centuries. In 1898 a lead weight, hanging from the spire down to the basement, was installed to measure the inclination.

Since 2003 there is an instrumental system that monitors the tower and the cathedral and is periodically upgraded (as illustrated in Chapter 5).

From a typological point of view, the tower recalls the Lombard bell tower model, with a square base of 11 m per side, which lightens upwards through the progressive increase in the number of windows. The tower exterior has five horizontal partitions consisting of arched cornices with decorated shelves that identify six different floors; however, this design does

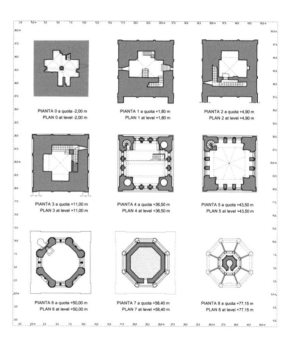

Figure 2.10 Horizontal sections at different levels

Source: Giandebiaggi *et al* (2009).

not correspond to the internal organization, and the path of the stairs interferes with some of the openings of the third and fourth floors.

Above the square tower shaft, there is a smaller octagonal drum and a high spire crowned by a golden copper sphere topped with a cross, for a total height of 89.32 m.

The structure is made of brick masonry cladded in slabs of stones of different thickness, sourced from Roman remains up to the fifth floor, while on the upper floors the material (ammonitico veronese limestone) was purchased specifically.

Twenty-one varieties of stones have been identified in the covering (Fig. 2.11), coming not only from Italy but also from Greece (cipollino marble) and from Turkey (proconnesian marble; Lugli *et al.*, 2009).

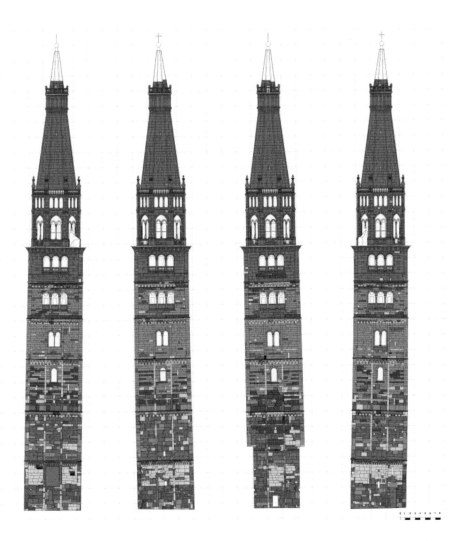

Figure 2.11 Map of the different lithotypes identified on the wall of the Ghirlandina Tower (Lugli *et al.*, 2009)

From the study of archaeological remains we know that these materials were imported by the Romans from distant countries to adorn the main monuments of ancient Mutina. Modena, which Cicero called *firmissima et splendidissima*, was in fact one of the most important Roman cities in northern Italy, and although there is no evidence of this glorious past, it is possible to picture the presence of the huge buildings from which large slabs of Istrian stone, in some instances finely decorated, were taken to be reused in the tower.

Although stone reuse was practiced also in the cathedral, the cladding is arranged in a different way, more regular in the church and more articulated in the tower, with blocks of different sizes and shapes and several decorated Roman pieces, which were partly removed during the restorations of the late 19th century when the shops that had been set against the tower were also eliminated (Fig. 2.12).

Even if the stone materials used for the cladding are the same, it has been noted that the Aurisina stone (in the varieties aurisina fiorita and granitello), an extremely resistant material, had a prevalent use in the tower, perhaps because, at least at first, the Vicenza stone was saved for the Duomo (Lugli *et al.*, 2018).

Figure 2.12 Replacement of stone element during the 1972 restorations showing how in some cases the cladding had become completely detached from the brick wall

Photo by Carlo Orlandini, *Raccolte fotografiche modenesi Giuseppe Panini.*

In addition to the stone material for the cladding, bricks were also salvaged, probably from portions of Roman buildings that, in the early stages of the construction, were still visible, while excavation (believed to be guided by divine indication[20]) was necessary in the later stages.

At the ground floor the walls are good quality and over 2 m thick, decreasing in thickness with height, down to 60 cm in the upper part of the spire.

The only entrance is the door opened on the southern side in the early 20th century to replace the original one that was now buried in the basement, no longer usable for a long time because of the settlement suffered by the tower.

Eight steps lead to the room used for reception, which has a ribbed cross-vaulted ceiling and two large corner pillars. There are two windows facing each other; the eastern one has a wider splay and a staircase. During restorations commissioned by the municipality in 2017, a door, whose trachyte lintel is visible, was identified on the north side just in front of the entrance. These entrances granted access from the small buildings that were at the time connected to the tower.

A staircase descends into the cruciform plan basement, with four large angular pillars of which the southwest one is larger in size because it contained the original staircase, no longer visible and of which we do not know any detail (Fig. 2.13).

At the center there is a masonry pillar that supports the cover with four irregular shape vaults. At the beginning of the 20th century the room was raised of 1.20 m[21] to avoid water infiltration so that now it has a very low height, and it is possible to closely observe the texture of the vaults and of the irregular arches. This anomaly perhaps comes from the position of the slit on the east side, now infilled due to the sinking that prevented a regular design.

Back to the entrance floor, the main staircase goes upstairs along the south side. This is the only room that still maintains frescoed[22] walls and ceiling, and it was used as a treasure room. The Ghirlandina is peculiar also because of the dual religious and civic role it played over the centuries: it is owned by the municipality, but still it is the bell tower of the cathedral. In the past it served important civic functions, as it housed the *Sacrestia communis*,[23] and from its top it was possible to overlook the city gates. This room is called *Sala della Secchia* (Fig. 2.14), a name that comes from a wooden and iron bucket that the Modenese people stole from a well in the city of Bologna during the Battle of Zappolino in 1325 and that was here kept as a war trophy. Specific information about this room can only be found starting from the 14th century, the period that also dates the decoration of the walls, which represents a

---

20 In 1167 the *massaro* of the cathedral had "the right and power to dig stones in the streets and squares of the city, as long as this does not inconvenience the inhabitants".

21 Infill intervention by Raffaele Rinaldi, alias "*il Menia*", estimated between 120 cm and 150 cm.

22 Agnolo e Bartolomeo degli Erri, *Coronation of the Virgin and Saints*, 1462–66, Modena; Correggio, *Our Lady and Saint Sebastian*, 1524 Dresden; Correggio, *Our Lady and Saint George*, 1530–32, Dresden; in these two paintings there are pinnacles and balconies on the balconies, while in the 1633 banner of the Commune of Modena by Lodovico Lana of 1633 the gables are no longer present.

23 Sacristy of the Commune: in 1327 the citizens' statutes mandated the archives of the commune to be kept in there. Here the treasure chests, public deeds and objects of high symbolic value such as the famous "stolen bucket" ("*la secchia rapita*") were kept. "*La Secchia Rapita*" was made famous by the heroicomic poem by the same name written by Alessandro Tassoni in 1622. The room had this function until 1578.

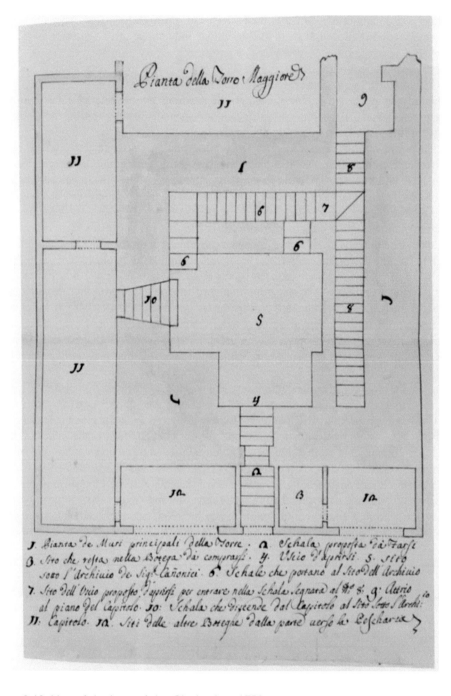

Figure 2.13 Map of the base of the Ghirlandina, 1750

Source: Modena, City Archive.

Figure 2.14 The Secchia room

vaio (a drapery made of fur), hence denoting its important civic function. On the vault there is a starry sky organized in a square mesh grid.

Under the plaster there are traces of a previous decoration, dating back to a previous construction phase (1184; Piccinini and Fiorini, 2015).

The room is divided into two superimposed chambers; the top one was smaller and located in a space above the main staircase and was reached by means of a ladder that no longer exists.

The public use of the premises gradually decreased in favor of the canons, and there were many disputes for its use until the State Property Administration finally declared it municipal property at the beginning of the 19th century.

Farther on the "well" of the tower, a single-room environment over 20 m tall, covered with vaulted ceilings, with large angular pillars supporting the flights of stairs. Along the walls there are windows that are mostly inaccessible due to the height; moreover, the single-light window on the north side and the two-light windows on the south side are crossed by the stairway at half height. This overlap could be the result of an afterthought in progress, or a modification of the staircase, or it may represent the choice of conditioning the internal arrangements to the exterior design.

The construction of this part could have happened in two phases, or simply have had a very long duration, because the inclination between the walls of the second and third string-course moldings is not homogeneous and thus demonstrates the correction of a settlement that occurred during the construction. A long time frame would also justify the presence of different manpower in the execution of the sculptures, clearly visible between the second and third moldings.

On the upper floor there is the *Sala dei Torresani*, the keepers of the tower (Fig. 2.15). This room had been separated from the walls to obtain a two-level apartment with a wooden deck that was removed by the municipality in 1925.[24] At the end of the 16th century, the residential area was reduced with the construction of a lookout towards the duke's castle.

Figure 2.15 The *Torresani* room

---

24 In 1598 the Este family was forced to transfer the capital of the dukedom from Ferrara to Modena and had the castle replaced with the current *Palazzo Ducale*.

Two stone benches were placed by the windows, and the community emblem[25] was painted higher on the wall.

There are four corner pillars that here are only partially hollow; three are rounded towards the center of the room to make space for a stone column with a cubic capital holding a shelf to support the tower keeper's apartment, later demolished. As evidence of this residential use, a chimney to heat the room that used the pillar space as a flue has recently been discovered in the southeast pillar.

On the perimeter walls there are four three-light windows, later partly reduced, which are separated by reemployment stone columns with eight capitals, dating back to well before the 13th century (Lomartire, 2015). Among these there are two with a clear narrative intent. The capital "of the judges" (Fig. 2.16) tells about good and bad justice; one side depicts a corrupt judge, kept on a leash by the devil, who takes money to give an unfair judgment, while on the other side, the honest judge is crowned by the angel. The "music" capital (Fig. 2.17) represents King David playing the zither, perhaps referring to the king's dance before the Ark of the Covenant.

Also, in this case, two stories of civic and religious nature represent the dual symbolic value of the tower. It is not known why these capitals, so different from the others, are in this room; what is certain is that the figures of the capital of the music have an explicit reference to the angular sculptures that decorate the third string-course molding, where we find the same characters depicted in a larger size.

For a while this was the last floor of the tower, with a flat roof, crowned with four corner turrets.

The next floor was built starting from 1261, which began the last construction phase. An articulated series of wooden structures supports four bells, today equipped with hammers that prevent the rotation during ringing that would provoke movement in the structure.

In the northwest pillar, which probably corresponded to one of the turrets, there is the spiral staircase leading to the bells room, where the structure still has hollow and round corner pillars on the inside and at the side has four large three-light windows closed at the base by stone balustrades. It is covered by a large vault called *lavezzo*, "basin" in the local dialect, meaning that the vault is similar to an inverted basin.

In the past, this room was consolidated from the inside with tie rods on all sides to counteract the tendency of the structure to open along the fissures that run throughout the upper part of the tower. These tie rods, however, do not connect to the external walls, only to the internal pillars.

The stone spiral staircase climbs to the last room, partially exiting the profile of the tower when emerging from the square part.

The spire was completed in 1319 but modified during the 16th century, also to repair the damage produced by a series of earthquakes that from 1501 had caused cracks to the spire and the drum. The intervention is still visible in the stone cladding placed on that occasion around the drum, which covers the upper part of the pointed arches of the mullioned windows (Fig. 2.18).

An impressive wooden helical staircase, built by the municipality in 1606, rises on the inclined walls and connects the two balconies, placed, respectively, at 60 and 78 m above

---

25 The coat of arms of the commune, painted on plaster, is represented by a shield with the cross, sided by with two drills, tools used to drill water wells.

Figure 2.16 The capital of the judges

Figure 2.17 The capital depicting the music

Figure 2.18 The Gothic mullioned windows covered by facing arches between 1575 and
1587

the ground. The lower balcony was decorated with *edicolette*, pinnacles and pierced gables, which are depicted in many *frescoes* (Fig. 2.19) and were probably removed during the 16th-century intervention.

From the balustrade of this balcony in May 2006 a large fragment of stone detached and fell onto the street below without causing damage, but this event started an important phase of study and the new restoration campaign.

2.4 The tower's sculptural apparatus

The façades of the Ghirlandina Tower are characterized by a significant sculptural apparatus, which, in the most ancient phase, is consistent with the design guidelines as the cathedral, although without equaling the quality level (Lomartire, 2015).

The five moldings have a total of 254 shelves decorated with animal or plant protomes, of which only 136 still carry decoration due to the substitutions with simpler geometric elements that took place in several instances. The first four floors are marked by pilasters that divide the walls into three parts and contain the windows but do not continue on the upper floors due to the presence of the three-light windows that are larger in size.

The moldings of the first three levels also have angular sculptures, with the exception of the first which on the south side is absent due to the connection with the arches of the cathedral.

As said, these moldings do not correspond to the real internal subdivision of the floors: the first molding runs at about half the height of the Secchia room, with intertwined arches, with

Figure 2.19 Correggio (1525) detail from *La Madonna di San Sebastiano*, Dresden.

Note: We see the tower still with the angular turrets and the triangular gables.

18 small shelves repeated on three sides, with human or animal protome, similar to those of the cathedral's apses (Fig. 2.20).

This molding carries five angular reliefs that depict a centaur in the act of hurling an arrow, a dog with a fishtail and a two-tailed siren sculpted on the same block, a feline carved in a Roman pillar of proconnesian marble (perhaps made by the school of Wiligelmo) and finally Samson, who smashes the lion's jaw (Fig. 2.21).

Figure 2.20 Detail of the cornice on the first level, east side

Figure 2.21 Sansone, first cornice, west side

The frames of the second, third and fourth level correspond to the free space called the "well" of the tower.

The second molding is a simple arch with a double ring; on each side there are 12 shelves larger in size than those on the level below. There are only six surviving protomes of the original 48, but they are sufficient to identify a different workforce from the previous one.

Just above the molding it is possible to recognize a correction of the verticality of the structure.

In the angular reliefs there are only depictions of animals: an eagle, two dogs and three lions, as well as an elephant and a calf now hardly readable.

The third molding marks another change in the characteristics of sculptures, intertwined arches return and the number of protomes is maintained, but the angular reliefs are larger than the others and depict human figures, in which similarities have been recognized with the ones of the parapet of the cathedral. It is unclear what they represent and if they have a single narrative intent, for example, that of music, but it is possible to recognize King David playing the zither (Fig. 2.22), a dancer, a woman with a flower, a couple holding hands perhaps in the act of dancing, a man with a shield and a sword and a man sacrificing a kid. The first three figures are the same that are depicted in a smaller size in the aforementioned "music" capital

Figure 2.22 Southeast corner sculptures of the third cornice depicting King David and a dancer

that is inside the *Torresani* floor, a floor that was built several years later. The protomes of this molding are executed with considerable skill and detail, especially in the representation of different pairs of animals. Also present are the symbols of the four evangelists, as well as different human and animal heads.

The fourth molding recalls the simple arches model and includes a sawtooth upper band. There are no angular sculptures. Among the 27 protomes there are no longer pairs of animals but masks drawn with marked traits prevail.

The main feature of this level is the presence of large mullioned windows, whose deep intrados houses paired columns on which are placed fine Granitello capitals (Fig. 2.23), which reproduce the Corinthian model with articulated variations, such as goats heads nibbling the leaves (west side) or with intertwined fruits (east side; Fig. 2.24). Only one capital differs from the others and depicts four eagles with spread wings, with their claws hooked on the collar at the base of the capital.

All these elements are the work of educated and highly skilled workers and carry the same formal connotations that we find in the capitals of the *Porta Regia* and are evidence of the permanence of contacts, or of a single activity, between the construction of the cathedral and the tower. The fifth molding is made of simple arches and has a sawtooth upper band.

The protomes are similar to those of the underlying molding. The level of the molding corresponds to the upper part of the *Torresani* room.

The three-light windows of this floor have semi-columns and capitals also on the external front, partly hidden by the masonry infill.

Figure 2.23 The capitals of the mullioned windows, west

Figure 2.24 The capitals of the mullioned windows, east

Figure 2.25 The animal figures and the human figures

The capitals recall the Corinthian model and the one that depicts leaves moving with the wind is particularly interesting. Just as inside, in the upper part the capitals have pediments decorated with models that are similar to those of the shelves.

At the sixth level, the floor of the bells, there are other large three-light windows with cubic capitals and balustrades with columns.

The square part ends with a thick and notched cornice and is covered by lead plates up to the perimeter of the octagonal drum.

In the last construction phase the complex decoration of the underlying floors is no longer present, the elements that adorned the balconies are lost and only the lower balustrade pinnacles rebuilt in the 17th century remain.

Chapter 3

Previous restoration work

Ancient modifications and 1970s' interventions

3.1 Introduction

Problems related to the leaning and the stability of the tower, arising from the interaction with the soil and the cathedral, as well as its seismic vulnerability, are addressed in Chapters 5, 6, 7 and in the Appendix, while the aim of this chapter is to reconsider all restoration and maintenance interventions since the 16th century.

At first glance the reader could have the impression that this content is rather a chronological collection of data, but deeper scrutiny reveals some important aspects.

First, the key to defining appropriate repair and restoration of built heritage is the awareness of the differences between modern and historic buildings. When dealing with built heritage, the history is rather complex and spans over centuries because of stages of construction, interruptions, modifications, rebuilding and repairs, all aspects to be investigated before making decisions. Therefore, such a reconstruction of events is of primary importance and is not only a matter of historical archiving.

In addition, it is essential to realize that today, most of historic building repairs are required sometimes as a result of the natural degradation of the fabric and the persistence of problems linked to the peculiar shape of the construction, as it is for the Ghirlandina Tower, which is very tall and isolated and therefore difficult to inspect and carry out maintenance.

But in many cases, repairs are also required as a result of damage caused by inappropriate repair over the last centuries, from incorrect mortar repair or expanding rusted iron in old stones or use of resins, causing spalling or delamination just to give some examples. For this reason, the preferred option is always minimal intervention and the use of traditional materials and techniques.

In this respect, reading this chapter will also provide the reader an awareness of changes that took place during the time in the philosophy of the repair and maintenance.

3.2 Ancient interventions and rebuildings

As described in Chapter 2, the construction of the tower lasted almost 220 years, and as it is quite common with historic structures, details of the first stages and time sequence of the construction of the oldest parts, as well as interventions that took place after the completion of the tower in 1319, still remain not fully known.

Most of the interventions listed at the end of this chapter refer to the pyramidal body and is work required to solve problems that are still "active", such as water infiltrations, weathering and stability of the stones.

The reason for the persistence of these problems relies on the shape of the tower, which is very tall and isolated and therefore difficult to inspect the surface and carry out maintenance, if not with adequate equipment and personnel, with consequent high costs.

The recurrent solution adopted over time has been the timely replacement of weathered stone slabs and different ways of grouting the joints between the slabs. Among the technicians who over the centuries took care of sealing the cracks there were essentially two different attitudes: some of them, according to an ancient technique, favored pouring lead in the cracks; some others considered this operation harmful because it broke the stones and therefore preferred specific stucco recipes, such as those mentioned in the list, or copper cladding.

To solve the more severe and extensive degradations, the external face of the spire was almost completely replaced twice.

3.3 The 16th century

The 16th-century substitution, carried out between 1575 and 1587, consisted in "jacketing" the entire octagonal portion, both drum and pyramid, adding a total height of 14 arms (about 7.32 m). The only elements of the previous arrangement still visible from the outside of the drum are the gothic arches of the two-light windows, which are covered in the upper part by the round arches of the new lining. The ancient structure, that of 1319, is no longer visible anywhere in the pyramid.

It may be possible to identify another trace of this intervention inside the space of the spire, climbing on the spiral staircase at about two-thirds of its height. The walls abruptly become thinner (as deduced from surveys,[1] the thickness is reducing from 90 cm to 60 cm) and slightly change in inclination, forming a recess that is present all around (see Fig. 3.1). This may be the point beyond which the old pyramid was demolished to make the new compartment obtained with the superelevation entirely usable and to install a staircase to reach the upper balcony. By graphically reconstructing the previous pyramid, it is possible to obtain about 14 superelevation arms, as mentioned in the documents (Fig. 3.2).

Recent radiocarbon dating analysis (see note 15 in Chapter 4), performed on a sample of mortar from the terminal part, presumed demolished, indicate an interval between the 15th and 17th centuries and thus include the proposed age. However, a definitive age could only be established through further research: once again, the need to unite scientific investigations and knowledge of the structures must be pointed out, even in the presence of historical documents, to avoid arbitrary interpretations.

On the other hand, there is an element that testifies the presence of the ancient portion in the lower part of the spire: a fragment of a 14th-century *fresco* found during the 2009 restorations inside the drum on the southeast side. This finding contrasts with the presence of the reinforcement "internal shell" identified by the experts who foresaw the 1890 restorations.[2]

During the intervention for the elevation of the spire, the covering was also replaced in other parts of the tower. In addition, the corner towers were demolished, probably because in a state of advanced degradation or because, being too close to the drum, they prevented the construction of the external covering.

---

1   The metric and photogrammetric survey was performed between February and July 2006 by the University of Parma with FO.A.R.T Ltd (Fotogrammetria Architettonica Rilievo Terrestre).
2   Serchia L., *Studi e interventi sulla Ghirlandina, i restauri*, p.176, in "I restauri del Duomo di Modena 1875–1984", edizioni Panini, Modena 1985

Figure 3.1 Staircase inside the pyramid

The two balconies were instead rebuilt at a higher elevation, while the one around the towers, clearly visible in paintings of the time,[3] was eliminated. There are also five reinforcement rings with tie rods in a sunburst arrangement, still existing.

3 See Niccolò Abati, The Adoration of the Magi, San Polo d'Enza parish church.

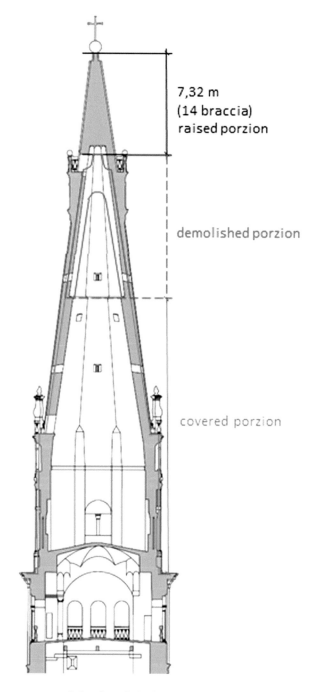

7,32 m
(14 braccia)
raised porzion

demolished porzion

covered porzion

Figure 3.2 Reconstruction of the demolished portion (in red)

The sphere and the cross were disassembled, cleaned, restored and finally put back on top and this initiative had great prominence: the chronicles[4] refer that on Friday, 19 June 1587, the shops stayed closed and the cross was carried in procession to the cathedral. All the civil and ecclesiastical authorities participated in the ceremony, the procession being accompanied by the sound of bells, drums and other instruments. Holy Mass was celebrated by the bishop and blessed with the cross. In the afternoon the cross was carried on the shoulders from the Duomo[5] up to the tower, the last stretch through the iron stairs outside the pyramid.

Other interventions were carried at the end of the 16th century, including the partial transformation of the *Stanza dei Torresani* (Room of the Tower Guards), into a belvedere, looking north towards the ducal castle, which was transformed a few years later into the still-visible *Palazzo Ducale*.[6]

3.4 The 17th century

At the beginning of the 17th century, a new phase of restoration began under the direction of the municipal architect Raffaele Rinaldi, known as *Il Menia*. It concerned mainly the consolidation of the structure by means of thickening the internal buttresses in the basement and at the bells floor. Slab detachments were also repaired on both shaft and pyramid walls, as well as the slope of the roof above the square part, because it had lowered towards the inside due to the weight of the spire.

Probably the large stone plumes in the lower balcony, of which there are the drawings of the wooden structure used to mount them, also belong to this phase.

3.5 The 19th century

Only in 1883 was an appraisal made of new important works. Although the Ministry of Education had asked to limit substitutions, safety reasons led to the replacement of the cladding slabs of the entire pyramid, perhaps with the only exception of the terminal section covered in lead. This section, the first to be restored in 1890, probably underwent the sole arrangement of the lead, while the integration of part of the stone cladding occurred earlier, as suggested by a stone arrangement which is completely different from the rest of the pyramid, alternating between narrow and wide courses (Fig. 3.3), and by the presence of 17th-century inscriptions on some slabs.

A wooden structure was built around the pyramid and cableway elevators were installed (Fig. 3.4) to operate all the cladding substitutions.

The second remake of the pyramid's stone cladding is also documented by a drawing by the municipal technical office, dated March 1897, which reported the lots and the dates of execution, the details of the ribs and the horizontal sections of the two balconies and of the *Torresani* floor (Fig. 3.5).

---

4 ASMo, sala ragioneria, relazioni e referti 1586–1587, vol.V, c.70v e segg. See O.Baracchi e C. Giovannini, "Il Duomo e la Torre di Modena", 1988, Aedes muratoriana, pp. 198–202.

5 At that time the access to the tower was made possible only through the cathedral.

6 In 1598 Modena became the capital of the Este dukedom and the new demands of the court led to the progressive transformation of the ancient and inadequate castle into the *Palazzo Ducale*, which, since 1859, was used as a military school and is still today the seat of the prestigious Military Academy.

Figure 3.3 Upper part of the spire: the arrangement of the stones according to bands of different thickness can be appreciated

The technique used to assemble the large Verona limestone slabs appears to be particularly accurate.

The slabs are arranged with the longer side on the horizontal plane, while the opposite happens in the upper row. The horizontal joints are internally covered and sealed with lead strips to prevent infiltrations and to ensure adherence; the ribs are part of the different slabs, not independent elements, and, like all other slabs, are alternated and fixed internally with metal pins, brought to light inside the pyramid during the 2009 restoration works.

The five orders of tie rods in sunburst arrangement, largely oxidized, were also repaired. The upper balcony was rebuilt, maintaining the parts that were still intact and covering the floor with lead.

Despite the accuracy of the project and the considerable expertise with which the work was carried out, after some time, water infiltrations recurred.

During the 20th century, the subsidence of the foundation and the movements of the stone cladding continued, creating new cracks and infiltrations. The degradation situation worsened due to atmospheric pollution that attacked the surface of the stones, as was the case in many other cities that underwent rapid industrialization like Modena.

3.6 20th-century interventions

Seventy years after the last work, facing the evident deterioration, the municipality planned a new important intervention and, in 1968, presented the project of restoration of the whole tower, with a purely conservative intent. The project proposed the use of new materials

Figure 3.4 Scaffoldings of the 1980 restoration works

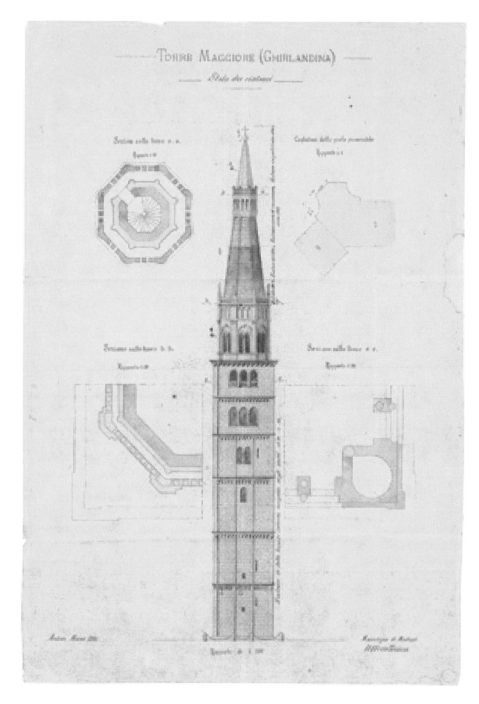

Figure 3.5 Status of restorations, as it appeared in the drawing of the technical office of the municipality in March 1987

Source: ASMo.

(resins) for sealing and surface protection, in addition to the replacement of degraded stones and other internal works.

The project report[7] concerning the exterior works reads:

> "removal, repair with possible replacement of parts of all parts tampered with moldings, cornices, inlays, etc., reconstruction of missing pieces and replacement of the same with the necessary clinging in accordance to best practices, using marbles of the same nature and origin of those removed, checking and testing the remaining parts, cement sealing of all cracks, complete cleaning of weeding and disinfestation treatment, water repellent protection of the surfaces, electrostatic protection".

And for the interior, it reads, "peeling and remaking of plaster, verification of the stability of the staircase ramps [. . .], re-painting, window frames".

In the verifications and comparisons necessary for the approval, several inspections of superintendence and experts, such as Prof. Cesare Gnudi,[8] were carried out to evaluate especially the interventions to be carried out on the sculptures.

Further checks were requested from the municipality on the products intended for grouting and consolidation, namely, vinyl-acrylic resin (CP5007) and epoxy resin (CP 415) from the company Sinmast. Initial tests were performed on samples in the lab, while there is no trace of further testing in the documents.

In the meantime, numerous sealings were made between the stones with an epoxy resin–based mixture (in some cases with the addition of grit and stone powder), slabs and moldings surfaces were treated with highly diluted single-component epoxy resins, and many slabs were replaced.

Because of time constraints and awaiting the opinions of the experts, the chief engineer[9] suggested that to differentiate among sculptures, only exhibiting relative weathering and be treated with diluted epoxy resin and those that were in a grave state of disrepair. The latter were replaced with a geometric looking slab of stone, postponing the intervention until the opinion of the Restoration Institute in Rome was formulated.

The restoration, performed with this technique, lasted for a long time, and in the following 35 years, only some interventions due to the separation of fragments on external parts were necessary, but no organic maintenance plan was foreseen.

The effects due to the use of epoxy resin, which, then, were not fully known, have been closely observed only in 2007 with the scaffolding installed for the new restoration works (see Cadignani and Valli, 2009).

The good resilience of the resin is in fact proportional to the resistance of the stone; that is, on hard and resistant stones such as *granitello* and *pietra d'Aurisina fiorita* there were no significant detachments, while on Vicenza stone, trachyte or *rosso ammonitico* the results were not acceptable because it caused detachments, sometimes massive, especially when the amount of resin used was significant. The phenomenon has been widely detected both on the slabs and on the frames (Fig. 3.6).

---

7 See Archive of the Modena Municipality, *Ripartizione Lavori pubblici*, Eng. Mario Pergetti, Prot. 8018/1972.
8 President of Center for Conservation of Outdoor Sculptures in Bologna.
9 Ugo Cavazzuti, chief engineer of Modena municipality.

Figure 3.6 Example of detachment on Vicenza stone treated with epoxy resins

Over the years the restoration techniques have changed considerably and the massive use of epoxy resins, which at the time was already discouraged, has been abandoned. They are still used only for the adhesion of detached stone fragments or in special cases.

A photographic campaign carried out during the restoration of the 1970s by Orlandini[10] has made it possible to identify many of the substituted elements and to see the masonry structure beneath the slabs that is not always well-bonded, also due to the numerous substitutions of the covering slabs.

The replacement of at least 22 protomes with new geometric elements. In the cornices, at least 70 slabs on the façades are documented. These numbers are to be added to the substitutions in past centuries, with a total loss of small decorated shelves of 48% compared to those originally present.

The now-"historic" photos show that at the time of restoration the surfaces were notably degraded, especially as a result of the effects of atmospheric pollutants and a degradation by a biological attack that have not been repeated so much in the following eras.

The adoption of methane-based heating systems in the city, especially after 1962,[11] brought about a significant improvement in air quality. However, due to climatic

10 See *Raccolte fotografiche modenesi*, Giuseppe Panini, Modena.
11 See A. Giuntini, G. Muzzioli, *Al servizio della città*, Bologna, Il Mulino, 2003; A. Giuntini, Il metano a Modena, in *Rapporto sulla situazione economica e sociale della provincia di Modena 2000*; P. Dogliani, *AMCM Energie per la città*, Modena, Ed Cooptip, 1987. V. Bulgarelli. *L'ambiente che quasi non si vede. Reti e servizi per l'energia e l'ambiente. La città e l'ambiente. Storia delle trasformazioni ambientali e urbane a Modena nel novecento*. Ed APM, Carpi, 2009.

conditions, significantly increased traffic and industrial emissions, the city today is in one of the most polluted regions in Europe and the monuments continue to suffer pollution damage.

The historical documents preserved in some of the most prestigious and ancient Italian archives, such as the Capitular Archives[12] and Modena's Municipal Historical Archives, allow only to hypothesize the execution methods of some of the interventions on the Tower after 1319.

It may be of interest to introduce a small digression on ASCMo, the Historical Archive of the Municipality of Modena. It is the most important documentary complex entrusted to a municipality in Emilia Romagna. It preserves a broad and almost-complete documentation of the political-administrative activity of Modena: first as a "free municipality"; then as a community of the Estense State, of which it became the capital in 1598, becoming particularly relevance; and then again as an Italian municipality after the unification of Italy. It is one of the few historical archives of municipalities of great tradition, the oldest part of which has not been deposited in the competent state archives. The headquarters of the archive, from the 14th to the 16th century, were in the Ghirlandina Tower itself. The archive preserves an imposing number of writings, maps, drawings and related to the testimonies of the numerous activities of the community.

Thanks to all the previously quoted documentation, today it is possible to reference numerous reports of damages and/or repair interventions,[13] even though not everything that was detailed in the chronicles was actually implemented. Those that are listed in the sequel refer to the implemented ones.

Chronological events and interventions of major concern:

1344 and 1345: damage was caused by lightning, which in 1347 caused the death of a tower guard.

Summer of 1481: once again due to lightning, the spire was urgently repaired to avoid substantial fall of stones.

1488: two arms long iron was purchased to secure the stone slabs.

1501: a ruinous earthquake caused damage to the ornaments, the windows and also to the cathedral vaults. The cathedral workshop commissioned the first indispensable works to Varagnana, but the works continued until 1530.

1524: lead flashings were placed over the moldings and the cornice.

1542: some pieces of stone fell from an ædicule, causing the death of a tower guard's wife.

1547: the municipality participated with the *fabbriceria* in the expenses to repair the spiral staircase that suffered from rot because of water infiltrations. The "flowered" tympanums were demolished.

1554: Paolo Castro, in charge of public works of the municipality, redacted a report on the tower's state of disrepair and mentions placing scaffoldings inside and out to put the "marbles" in place. The repairs were actually carried out within 1556.

---

12 It preserves the documents produced or received by the cathedral of Modena from the Lombard period to the present. The most precious collection is constituted by manuscripts that have come to Modena over the centuries

13 The quoted dates are mostly derived from the following texts: C. Dieghi, relazione storica, in Comune di Modena, Progetto di restauro della torre Ghirlandina, progetto definitivo, relazione illustrativa, febbraio 2007 and Fonti e studi per la storia della Ghirlandina in *La torre Ghirlandina, un progetto per la conservazione*, Sossella editore, 2009; O. Baracchi and C. Giovannini in *Il Duomo e la Torre di Modena*, 1988, Aedes muratoriana; C. Acidini Luhinat, L. Serchia, S. Piconi, *I restauri del duomo di Modena 1875–1984* edizioni Panini Modena, 1985.

1565: the tower guards asked the council to repair a room in their quarters.

1572: a commission was nominated to take care of the tower repairs.

1574: the Corporations of Art and the municipality contributed to the expenses for the tower's repairs.

1575: the new "marble" covering of the pyramid was planned, and its delivery was arranged by river transportation from Verona passing through Ferrara. Lumber was purchased for the tower scaffolding.

1580: due to the high expenses, the municipality proposed to cover the top part with lead.

1582: stones purchased for the tower arrive in the dock of Modena.

1584: the pyramid is elevated of 14 arms.

1587, June 19: the restored cross was blessed by the bishop and placed on top of the tower.

1606: the new wooden helicoidal staircase was constructed inside the spire.

1607: an excavation was started to verify the stability of the tower.

1609: Il Menia compiles a report in which he expresses concern for the tower's stability and proposes to thicken the internal buttresses to stabilize it.

1640, September 24: the top of the square part of the tower was covered with lead.

1651, 1658 and 1659: lightning damage was recorded.

1666–1690: a "stucco" was applied (composed of lime, marble powder, iron bitumen, stone powder, all mixed with walnut oil) all over the tower to close the fissures that caused water infiltrations, under the supervision of the architect Marco Costa.

1717: the "stucco" was applied once again to seal the stone slabs.

1733: repair works are resumed and the architect G. B. Massari adopts a new way of using lead as a covering and for repairs between the stones.

1765: the shops at the foot of the tower were demolished and the covering was integrated to close the niches.

1781: the Scarabelli–Pedoca report mentioned once again the problem of water infiltration in the pyramid and the deterioration of the covering stones and of the sealings due to grass growing on them, and lead payments are recorded up to 1890.

1794: architect Giuseppe Soli deemed the grouting unnecessary and proposes a copper covering of the pyramid.

1802: both Dondi[14] and Valdrighi[15] denounce a collapse risk, but the architect Soli, sent to check, certifies that even though the danger of collapse was nonexistent, the tower suffered from severe infiltration and it was necessary to intervene.

1807: engineer Blosi did an appraisal for the Demanio, where he detected disconnections and gaps in the stone covering of them in the pyramid, and he pointed out the need to redo the spiral staircase and to repair the one of the octagons. In the same year, poplar planks and oak trusses for the staircase were put in place.

1810: engineer Manetti, to avoid infiltrations in the pyramid, suggests that instead of copper, the *Podestà* to use a cement composed of pulverized roof tiles fired in a furnace, mixed with well-cleaned river sand and pig fat.

1813, March 17: Manetti inspected the tower and did not detect alterations in the robustness of the building, but he did observe deterioration in the form of scaling and foliation on

---

14 A. Dondi, canon and Modenese historian, author, among other things, of *Notizie storiche ed artistiche del Duomo di Modena, tip. Immacolata concezione*, 1896 Modena.

15 F. Valdrighi, Modenese historian, author of *La torre maggiore di Modena, tip. Sociale*, 1876 Modena.

the stone arches of the big windows. On July 20 he compiled a report summarizing all the interventions from the 16th century onward and criticizing the use of lead.

1828: the new lapidary museum was opened. On that occasion the decorated and inscribed "marbles" located in the lower part of the tower façades, were replaced with flat slabs and the original moved to the museum.

1842: repair work was performed on the northwest corner.

1847: an earthquake reopened the fissures in the bells room.

1883: an appraisal for the consolidation and ample substitution of the stone cladding of the spire was done and then sent for funding to the Ministry of Education, which rejected it in 1888.

1890: a new project was presented and the works were carried out between 1890 and 1897.

1889 and 1901: assessments on stability were conducted by Prof. Cavani and Prof. Canevazzi.

1968: a new project of global restoration was presented to the Ministry of Education[16] and then carried out between 1972 and 1973 with the requested modifications.

16 The Ministry of Education was at the time competent on Antiquities and Fine Arts, the Ministry of Cultural and Environmental Heritage was established in Italy in 1974 with the Legislative Decree n.657.

Chapter 4

Planning new investigation studies and restoration works

After the 1970s' restoration, described in the previous Chapter 3, the tower underwent only sporadic intervention until 2002, when small fragments detached from the external surface.

At that time, it was necessary to rely on climbing techniques to get an immediate visual inspection as well as to remove small portions of detached material and to detect the main lesions.

In October 2003, a monitoring system to record the tower's movements was also installed, and this system is still in operation today.

A maintenance work program was then set up and in February 2006 a metric and photo-grammetric survey, to serve as a basis for the project, was available. Moreover, the surface was again checked through the previously mentioned mountaineering technique, revealing a significant increase in the number of detachments in just four years. In particular, the largest number of damaged elements were found to be on the tower's east side, but the ones that were *more severely* damaged (though fewer in number) were on the south side

In May 2006, few months after this quick check and removal intervention, a stone fragment larger than the previous ones detached from the balcony at a height of 60 m and fell onto the narrow street below. This event, which represented a danger for the conservation of the monument as well as for the safety of passersby, induced the municipal administration to take more incisive and comprehensive measures.

Therefore, more than 30 years after the previous intervention, the municipality of Modena promoted a new project for the conservation of its iconic monument, which also resulted in the restoration of the spire.

This chapter deals with this project and its content is rather wide and specialized.

As it is often the case, cleaning of architectural surfaces was the main procedure carried out in the restoration of the tower, and it is shown how cleaning is rather delicate work because it involves largely irreversible operations. Therefore, the advantages and the short-comings of different cleaning methods, based on mechanical, physical and chemical procedures, are presented in detail.

In making decisions, it was of paramount importance to focus on the astonishing large variety of stones covering the surface of the tower, mostly recovered from Roman buildings (*spolia*) and reused. Therefore, this chapter deals particularly with this topic, as well the petrographic characterization of the mortars and plasters inside the tower, which allowed the identification of seven groups of mortars used in various construction phases.

Finally, the spire appeared to be the most damaged part due to water infiltration causing plaster detachment, instability of the wooden staircase and loss of stability of the balustrade elements, all challenging aspects that may be of great interest for the reader involved in the preservation of built heritage.

4.1 Investigations

The origin of the Ghirlandina Tower and the story of its modifications and additions not only provide a general idea of its construction development but also show how the study of ancient monuments can often be a long and never completely accomplished pursuit and how indispensable is the contribution of different fields in order to achieve a complete vision of their history.

For this reason, the new restoration was conducted with the contribution of a multidisciplinary team of experts, brought together in a Scientific Committee, appointed in 2007.

In the initial phase of the intervention a large scaffolding was installed, covered with sheeting decorated by the artist Mimmo Paladino.[1] In this way, the Ghirlandina Tower also became an example of public art for the duration of restoration works. On the large white canvas (see Fig. 4.1), the artist represented color geometries and black-and-white photographic fragments of his sculptures, as a dialogue between the art of the past and the present (Vettese, 2009), so effective that many citizens thought that the images represented the sculptures on the Duomo. The work, even if temporary, incited a lively debate in Modena between those who considered the cover an aggression to the monument and those who defended it as an artistic opportunity.

Only after the installation of the scaffolding was it possible to get a closer look at the surface of the monument and assess its status in detail.

This allowed studies in different areas and developed on the basis of agreements and research program with many universities.

4.1.1 Research focused on stones, bricks, mortars, plasters and wall paintings

It is rather astonishing that 21 varieties of different stones were identified on the surface of the tower, mostly recovered from Roman buildings (*spolia*) and reused. They come from other parts of Italy as well as from other countries: Istrian stone from the area of Rovinj; proconesium marble from Turkey; naxian, parian and cipollino marble from Greece; Aurisina stone from the Trieste area; rosso ammonitico and scaglia rossa from the Verona area; Vicenza stone from the Berici hills; and trachyte, the only igneous rock present in the tower, from the Euganean Hills near Padua.

In substitution interventions sandstones from Apenninic quarries not far from the city of Modena, travertine from Lazio and Carrara marble, biancone and bronzetto from Verona and Chiampo stone from the area of Vicenza were also used.

The rocks are of various origin, age and sourcing. Paleontological analysis was performed during the restoration campaign to identify the fossils (Papazzoni *et al.*, 2010) visible on the exterior of the tower. In particular, the identification of fossils and microfossils allowed to identify with certainty the lithotypes and the respective area of origin.

---

1 Italian painter and sculptor, one of the most famous exponents of Transavantguarde, born in 1948 in Paduli (Benevento, Italy). His works are included in important permanent collections in Italy and abroad. He made installations and interventions on urban spaces, such as the *Hortus Conclusus* in Benevento or the *Mountain of Salt* in Naples. In 2007 the municipality of Modena appointed him to create the scaffold sheeting for the restoration work. The choice fell on him because of his decades-long relationship with the cultural scene of the city and for his ability to interact with the public on the streets. The "canvas" was installed on 19 January 2008 and removed on 21 September 2011.

Figure 4.1 The scaffolding of the Ghirlandina Tower covered by a cloth decorated by the artist Mimmo Paladino

After cleaning, the walls turned out to be a proper outdoor paleontological museum, visible to the naked eye thanks to the temporary presence of the scaffolding. A catalog containing photographic images of hundreds of specimens was compiled and subdivided into eight main groups.

Rosso ammonitico (Fig. 4.2) and scaglia rossa show numerous macrofossils such as belemnites, ammonites and sea urchins. In Aurisina stone and Aurisina Fiorita, the characteristic

Figure 4.2 Ammonite section in the rosso ammonitico with the shell subdivided by sectors

fossils are rudists, while in the Vicenza stone, sea urchins, rhodoliths and coral colonies are visible.

Chemical and physical analyses carried out on samples of the four more common stone materials, that is, rosso ammonitico, Vicenza stone, Aurisina stone and trachyte, allowed the assessment of the state of degradation and the identification their characteristics.

Types of degradation were mapped according to their intensity, as shown in Figure 4.3 and 4.5 (Lugli *et al.*, 2009).

Classified according to standards, the most recurrent degradation phenomena were chromatic alterations, biological colonization, dripping, black crusts, differential degradation, surface deposits, disintegration, detachment, erosion, exfoliation, fractures, gaps, lesions, films and sulfation.

Black crusts were mainly affecting the south and west façades, toward which the tower is inclined, while patinas and biological colonization were more prevalent on the north façade, which receives less sunlight and therefore remains moist for longer. The other degradation phenomena are distributed on all exterior walls, with a modest accentuation on the lower floors on the north and west sides.

The nature of biodeteriogenic degradation was analyzed and identified with blue algae patinas, such as *Phormidium foveolarum* and *Microcystis viridis*, and the colonization of mosses and lichens, such as *Verrucaria lecideoides*.

It is important to note that artificial patinas, still partially present on the stones, were also analyzed. These are oxalates, derived from previous interventions performed using natural organic substances, such as caseins and animal glues, which, over time, not only assumed an

Figure 4.3 Stone deterioration mapping: black crust and soiling (Lugli *et al.*, 2009)

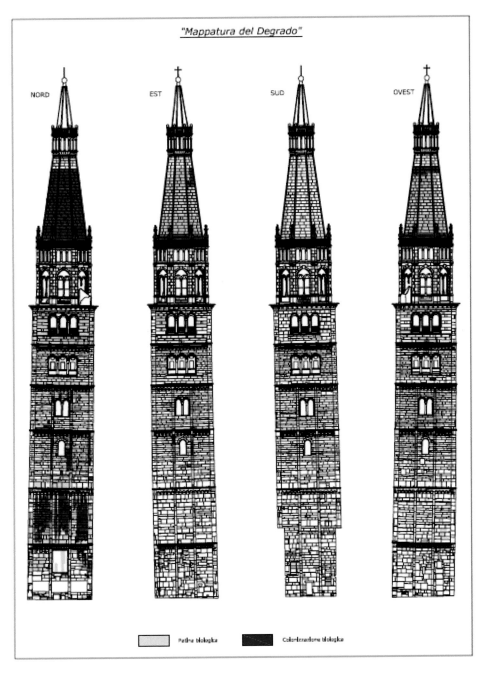

Figure 4.4 Mapping showing biological colonization and biological patina (Lugli *et al.*, 2009)

evident ocher/yellowish color but also kept the stone surfaces almost intact. In the areas of greater rainfall, these patinas were no longer present.

Last, a specific study was started on the artificial resins used in the most recent restorations to verify if it was possible to remove or at least to reduce them.

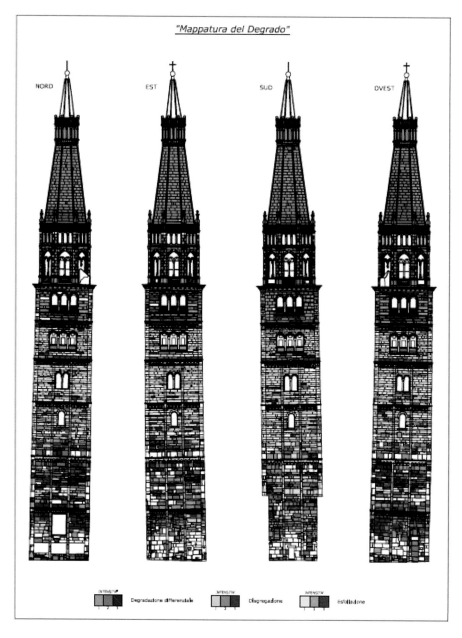

Figure 4.5 Mapping showing the differential degradation, disaggregation and exfoliation

Source: Lugli *et al.* (2009).

The set of investigations outlined a precise picture of the nature of materials degradation, confirming the presence of stone alteration phenomena that are typical of buildings in urban environments. It should be noted that, probably due to the widespread presence of oxalate patinas, the phenomenon of surface sulfation was not particularly remarkable. Where instead these patinas were no longer present, stones exhibited a larger extension of disintegration and gaps (see Figures 4.6, 4.7 and 4.8).

Stone elements and bricks inside the tower were also mapped, and this mapping allowed the identification of the stones used in lintels, columns and stairs (Lugli *et al.*, 2010). The materials used in the flooring and the Roman or medieval reused brick material on all floors were also surveyed. This study was matched with the archeological investigation on the construction materials of the elevations, which allowed to identify five construction phases and three modification episodes (Labate, 2010).

To capture the characteristics of the construction, in addition to main materials (see Colla et al, 2010), it is also important to identify the apparently minor materials used in mortars and finishings that may contribute decisively to define construction phases and methods.

For this reason, a petrographic characterization of the mortars and plasters inside the tower was performed, which allowed the identification of seven groups of mortars used in various construction phases, distinguishable by composition and grain-size distribution of the aggregate and for the characteristics of firing residues and lime lumps.

The most significant parameter for the classification was found to be the origin of sand aggregate within lime, from both the Secchia and Panaro Rivers (which flow, respectively, to

Figure 4.6 Black crust on the sculpture of the third cornice on the south façade

Figure 4.7 Pronounced biological colonization and biological patina on the third cornice of the northern façade. Notice how moss grows is more on the *pietra di Vicenza* arch on the left.

Figure 4.8 Example of differential degradation of rosso ammonitico, southern facade

the west and east of the city) and from smaller watercourses in the surrounding area (Lugli et al., 2010).

In addition, mortars were and are currently being analyzed using innovative survey techniques in order to date them. Radiocarbon analysis does in fact allow the dating of the carbon dioxide absorbed by lime and the determining the time of the construction of the walls (Lubritto et al., 2015).

As for the wall paintings, the room located on the first floor, called *Sala della Secchia*, is the only one whose walls and groined vault are completely frescoed. Micro samples of the main colors used were taken and analyzed (Baraldi, 2010), by using three different techniques: Raman microscopy to investigate the molecular nature of pigments and dyes, Fourier Transform infrared spectroscopy (complementary to Raman) and, in some cases, X-ray fluorescence spectrometry.

The analysis showed the presence of calcite in all samples, allowing to identify it as lime paint, applied, in some cases, with *fresco* technique and, in some cases, with *fresco secco*.

Strong sulfation of calcite and the presence of calcium oxalates caused by biological attack indicate pollution-related deterioration.

There are no materials that allow dating the paintings, if not those used in restorations. For example, *terra verde* is a pigment from Mount Baldo, used since Roman times. An iconographic historical analysis has ascribed the paintings to the 14th century for the typical upholstery design covering the walls, consistent with the use at the time for decorating main domestic environments with fine fabrics. The underlying plaster shows traces of an older decoration also painted in bright colors.

Once the nature and conditions of materials deterioration were determined, intervention techniques and the materials to be used were analyzed, favoring the use of reversible and low-environmental impact products and interventions, as well as lower levels of toxicity for operators and ease of disposal. It was therefore decided to use mainly water-based products, limiting as much as possible solvent-based products but testing them equally for case-specific use.

4.1.2 Detecting pattern of fractures

In addition to the discussed investigation on the surfaces of the tower, many efforts were devoted to geotechnical and structural aspects, which are described in detail in the following Chapters 5 and 7.

In this respect, studies were also carried out on the construction phases and the related corrections of the verticality of the tower, which was experiencing tilting since the first phases of construction, as it can be inferred from Figures 2.9 and 2.10 (in Chapter 2).

It was observed that the main lesions were arranged vertically and mostly located in the upper part of the tower, between 30 m and 50 m, in correspondence with the window openings (Alfieri et al., 2009).

Although chronicles of the 19th-century restorations report that the greatest damage was found on the north side, today injuries are much more evident in the south façade.

Lesions called "hourglass fractures" (see Fig. 4.9 and 4.10), from the shape of stone portions detaching at the corners of pilasters, were also detected in the external covering. This is a quite widespread type of lesion in all structures that have large masonry walls since mortar joints tend to shrink while stone cladding is stiffer.

Figure 4.9 Hourglass-shaped crack on a half pillar, with material expulsion. The joint between the two stones is nearly invisible in the central part of the hourglass.

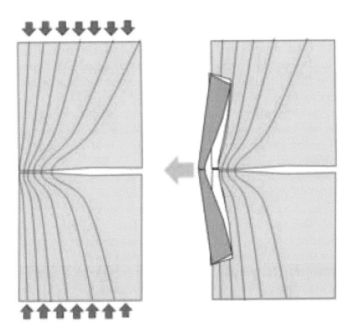

Figure 4.10 Scheme of the hourglass-shaped crack phenomenon present on some pilasters of the tower

Source: Alfieri *et al.* (2009).

The cause of this is the concentration of tensions between slabs that are juxtaposed with minimal joints and are stiffer than the internal masonry. The phenomenon had already been tackled during a previous restoration by plastering most of the lesions with resin. However, no significant results were obtained with this technique because most fractures are still active.

4.2 The diagnostic project

After the investigation on materials and degradation, a second phase of experimental studies was aimed at choosing the most suitable techniques and products to be used for the main types of degradation (Biscontin, 2009; Morabito *et al.*, 2009).

4.2.1 Cleaning and restoration tests

As it is often the case, cleaning of architectural surfaces was the main procedure carried out in the restoration of the tower. It was devised to remove deposits of harmful extraneous substances, such as atmospheric particle, soluble salts and combustion residues. Cleaning is the most delicate phase because it involves largely irreversible operations. The system used must therefore be well controllable to ensure a gradual and selective removal of deposits, it must be adapted to the specific conditions of degradation, and at the same time, it must protect the noble patinas present on the stones. For this reason, numerous cleaning methods, using mechanical, physical and chemical procedures, were tested.[2]

Tests with steamjet[3] technique gave significant results on a good part of the surfaces, removing the thick black crusts with the aid of sponges or soft brushes. In cases where thickness or adhesion was greater, it was necessary to back the technique up with a surfactant (triethanolamine). This method has a modest environmental impact as the use of water is reduced compared to direct use of nebulized water.

Other systems were tested, for example, based on urea-glycerin, a system that is more delicate than the one based on ammonium-carbonate but that has given good results only after a series of repeated applications. This technique would have lengthened the time of intervention too much and was therefore not adopted.

Tests were also conducted on a biodeteriogenic technique, which consists in applying a preparation based on sulfate-reducing bacteria, capable of attacking the black crust, in a period ranging from 12 h to 36 h. Effectiveness was guaranteed only in the presence of black crusts and not of other complex deposits. It is an effective and highly selective technique even on disaggregated supports, but it was not taken into consideration due to the difficulty in recognizing the deposits actually treatable and for the high costs.

Sometimes steamjet technique and carbonate wrap crust removal tests did not give good results, in some instances due to the excessive thickness of the crust, in others due to the presence of previous surface treatments.

As for the decorated surfaces, such as sculptures, where very cautious intervention is required, laser technology was tested and proved to be effective and delicate, even on the most compromised surfaces. Scanning Electron Microscopy (SEM) analysis showed that laser cleaning method did not damage the stone surface and allowed the preservation of the old oxalate patinas.

At the end of the tests, it was decided to intervene on all surfaces for the removal of deposits and crusts with the steamjet technique, integrated each time with other previously identified techniques depending on the case (Fig. 4.11).

Four biocidal products were tested for removing agents responsible for biological attack, which was more severe on the north side and in areas without oxalate patinas. The effectiveness and compatibility with stone surfaces were checked and a quaternary ammonium salt product that removed the biodeteriogens without being too aggressive on the surface was selected.

The product choice for the re-aggregation and protection of surfaces also followed the logic of favoring water-based systems, but solvent-based products were tested on site as well.

Five products were tested, taking into consideration the different types of nanodispersion (acrylics, silicons) and comparing them with solvent-based products. The products were left on site and evaluated for a seasonal cycle, verifying, in addition to the capacity for surface reaggregation and water repellency, as well as chromatic variation (color measurement). The results of the tests indicate that water-based products have the same efficiency of solvent-based products and do not cause changes to the aesthetic appearance of the stone.

---

2 Examples of mechanical methods are brushes and scalpels; physical-mechanical methods include micro-sandblasting and vibro-incision; chemical methods use carbonate poultices and enzymes; physical-chemical methods include nebulized water, Jos, steamjet; the physical method uses laser.
3 The steamjet method uses water that is converted into low pressure steam and reaches stone surface at low temperature.

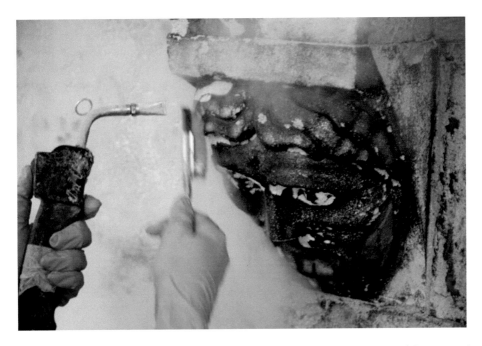

Figure 4.11 Example of vapor-jet cleaning of a sculpture placed in the second frame on the
south side

To solve the problem of rainwater infiltration between the vertical joints of the spire slabs, different ways of sealing the joints were tested. Among these, lead wire sealing was considered because of the presence of this material in the horizontal joints. This technique consists in compressing a thin lead filament into the cracks by means of a metal tip. The discontinuities detected with the optical microscopy observations, however, excluded the use of this technique.

Another test was performed on a system consisting of plastering the joints with a mortar based on fluorinated elastomers, which has good physical–mechanical compatibility with stone, excellent elasticity and a good degree of reversibility. Two products were tested, and their adhesion characteristics were verified. The compound was made of elastomer added with acetone and stone powder, in a ratio of 1:2. Plastering was performed in the deep part with lime mortar and in the upper part with the elastomer. From an executive point of view, a certain difficulty of use due to fast drying must be observed.

A specific investigation was aimed at removing mixtures of epoxy resin binders, widely used in the 1970s' restoration. Over time, the adhesive properties of the resin determined a strong interaction with the stone, deepening the detachments in the stone itself, resulting in the material being affected for a considerable thickness. This is a fairly new and difficult problem to solve because epoxy resin is considered very tough and almost irreversible. Laboratory tests allow the identifying a solvent system able to weaken the structure of the epoxy binder to ease its removal. This is a poultice consisting of a solvent mixture supported by a cellulose-based gel that can be applied to the resin and sealed with a plastic film. Tests indicate that effectiveness is obtained after about 7 d to 8 d from the application but weakening remains confined to the external part of the epoxy mix, which can then be removed by means of a spatula, and it is then necessary to repeat the operation a number of times depending on the thickness being removed. The use was very limited because it is a rather laborious operation.

During the diagnostic phase, a portion of a balustrade was dismantled to check the conditions of the individual elements and the construction technologies. The two balconies are protruding elements and therefore particularly vulnerable and constitute one of the main problems not only for the stone degradation but also for the static instability. The lower balustrade, located at 60 m, consists of 64 columns, 8 on each side installed on a single stone basement and surmounted by a cornice. At corners there are eight pillars that end with a sphere crowned by a flame. The upper balcony, at 78 m, is composed of 24 columns, 3 on each side, mounted on single dice, with cymatium and angular pillars decorated with spheres.

When checking the fixing systems between the elements and their state of degradation, it was found that balconies were the most degraded elements of the entire tower due to the large number of fractured elements and disconnection of joints.

Every single element composing the balustrades was therefore analyzed, also with an ultrasonic technique, to verify the characteristics of the material and to examine the degree of homogeneity and resistance. Many columns were so degraded that they did not allow consolidation to guarantee stability over time. Following a detailed petrographic study, the lithological characteristics and the paleontological content of the original rocks were identified, and it was possible to find the most suitable material to replace the columns that were no longer usable (Lugli *et al.*, 2009; see Fig. 4.12).

The set of results of the numerous series of tests, developed according to the objectives established by the committee, allowed the obtaining a good knowledge of material compatibility and improved the planning of a focused restoration.

Figure 4.12 State of decay of many columns of the balconies

4.3 The restoration

Restoration work took place from November 2008 to September 2011 and was divided into two successive phases, depending on grants availability. It considered first the upper part of the tower, starting from the inside and continuing outside in the warmer months.

4.3.1 The spire

The spire appeared to be the most damaged part, inside due to water infiltration causing plaster detachment and instability of the wooden staircase and outside due to a loss of stability of the balustrade elements (Fig. 4.13).

From a geometrical point of view, the spire is an octagonal pyramid with a total internal height of over 28 m. Attached to the inclined walls there is a helical wooden staircase that allows access to the balconies.

The intervention included the removal of old repairs carried out with cement plaster, during which an ancient *fresco* was found, identified by experts as belonging to the first construction phase. Various stratigraphic investigations on the walls identified a white lime wash over the original plaster, then a pale ocher, and finally a more recent sequence of darker ocher tints.

The flooring of the room follows the course of the bowl-bottom-shaped great vault, and over time, materials had been gathered on the sides that had changed the original level until closing the gutters[4] in the floor and incorporating a large part of the first step of the wooden staircase. To restore the original levels, the added material was eliminated, and a new *cocciopesto* flooring was laid over the existing one, which was very damaged.

All wooden window frames were restored with a synthetic wax–based product on the outside and a natural wax on the inside. After a few years, the treatment presented some cracks that made the protective effect ineffective.

The restoration in the external parts began with preliminary cleaning operations, the entire stone surface was dusted, then a biocide product was applied and the surface was washed, repeating the operation twice for complete removal. The cleaning of the concretions was performed with the use of the steamjet, integrated where necessary with other tested techniques. The golden sphere and the cross were also cleaned and their connections verified.

The most challenging intervention was the material and structural restoration of the staircase,[5] a wooden helical structure, with a parapet, supported by 22 thin metal shelves. It is made of different woods, mainly spruce, with structural parts of oak and poplar. In the past, the material was attacked by *Nicobium castaneum*, a woodworm typical of the Mediterranean areas that grows on damp woods in contact with the walls. The insects caused numerous flickering holes, but no live individuals were detected. The staircase was treated with an insecticide (permethrin) in early spring to eliminate any surviving individuals and to avoid new proliferations.

Cleaning was carried out with compresses of green clay on the floor and with paint-stripping agents through tissue paper on the balustrade.

---

4 There are no window fixtures inside the spire; therefore, water coming from the windows can be drained through gutters.

5 Structural consolidation project by Francesco Parolari.

Figure 4.13 Detail of the upper balcony closed off by wire mesh since 2006

The steps were reinforced, where necessary, with blocks of seasoned poplar wood, restoring stability between risers and treads. To reinforce the damaged load-bearing beams, new seasoned oak beams were inserted in parallel, maintaining the same wood essences as the original ones for each element and fixing them with wooden dowels. At the end of the consolidation operations, natural beeswax dissolved in turpentine with the addition of plants essences was laid on the surface, to help disinfection and conservation.

Analysis on the staircase-support brackets, a strut-and-tie system, showed the great slenderness of struts, a strut-and-tie-rod meeting point very far from the point of application of the joists and a welded connecting system with many fissures. To ensure greater stability, an integrative support system independent from the staircase was designed, supporting the existing brackets in correspondence of the rings that carry the longitudinal beams of the staircase. The strut-and-tie plates of the new shelves are fixed to the wall structure with threaded bars and can be removed without damaging the staircase. The new structure was designed to have a capacity of carrying 2 kN m (Fig. 4.14).

After this phase substantial consolidation work was started on the balustrades of the balconies. Disassembling tests had shown the presence of numerous fractures and disconnections that had to be repaired with fiberglass couplings and gluing.[6] When it was not possible to consolidate the columns safely, replacement was necessary. Thanks to the information gathered during the knowledge phase, a series of sampling and comparison surveys were carried out in various rosso ammonitico quarries in the Verona area and a block with similar but better characteristics than that of the columns to be replaced was selected (Fig. 4.15). The new columns[7] were shaped using numerically controlled machines on the sample of one of the oldest columns still in good condition. Eight columns were replaced in the upper balcony and 12 in the lower one (Fig. 4.16).

The fixing pins of the columns at the base and the cymatium, generally damaged and strongly disconnected, were replaced with new copper pins. Lead plates were relocated between the various parts of the balustrade, column and base, spheres and cymatium, and the joints were sealed with fiber-reinforced mortar. The lead-lined parts of the balcony floors were replaced with similar material using a similar technique. In particular, the lining of the lower balcony was completely detached from the wall, therefore allowing water to infiltrate the wall. The new material was inserted by repeating the same procedures, but after only three years the same defect occurred: the plates must in fact be anchored to a support structure to avoid that prevent them from moving from the seat in which they are housed with temperature-induced changes in expansion and movements between stones.

To prevent infiltrations, the rainwater removal system was reestablished by reopening of drainage holes and by improving the drainage systems that had clogged over time.

Elastomeric grouting was used in the spire to seal the joints between the stones. This material was chosen for its high impermeability and its ability to adapt to the support. During the intervention, after about 1 y from the application, however, a significant color variation was detected due to the tendency to absorb atmospheric particle deposits, and for this reason, the use of elastomeric grouting in the subsequent portions was abandoned.

4.3.2 The shaft

The second phase works involved the replacement of the lead covering of the square-based portion of the tower, very damaged and in which the different repair tests had yielded unsatisfactory results. During the removal of the slabs, in the southwest corner of the roof, a small slab of engraved lead was found, dated 24 September 1640, which reported the names of the

---

6   Gluing with bi-component resin.
7   The stone block is from the Cave Bonaldi quarries in Sant'Ambrogio di Valpolicella (Verona, Italy).

Figure 4.14 The wooden helical staircase after the restoration and structural reinforcement

Figure 4.15 Comparison between a fragment of the original columns and the block of rosso ammonitico selected for the preparation of the copies

clients and the workers who did the roofing. The slab was removed and kept at the city civic museum (Fig. 4.17). The text reads:[8]

> Adi 24 setembre 1640. Questo cornisone coperto da noi sotiscriti d(e) comisione di(l) signor Marcelo Spacini, dil signor Giminian Scalabr(in)i, dil sig(nor) Cesar Carandini, dil signor Iacomo Roncalia deputati dalla illlustrisima Comunità di Modena, et Frabbriciero (d)i detta tore et per sopra(s)ta(n)te *sopra a (d)etto lavoriero fu il Magnifico Cristofero Malagola detto Galaverna, io Giuliano Negri milaneso di una tera ciamata Meno rivera do(r)ta coperse detta opera con Paulo Si*ncia modenes compagni tutti doi.*

All surfaces, to the ground, were cleaned with the same method used for the spire: first they were dusted, then a biocide treatment was repeated and rinsed twice and then the deposits and black crusts were removed with steamjet technique.

Removal of the crusts from the shaped cornice that crowns the shaft was performed using compresses of ammonium carbonate in addition to the operations common to the entire surface due to the thickness and particularly strong adhesion of the crust.

The rich decorative apparatus is present in the cornices of the first five floors, which have decorated shelves, and in the double and triple windows that are adorned with capitals.

---

8 Text transcription by P. Bonaccini (University of Bologna).

Figure 4.16 The preparation of copies of the columns using a computer controlled machine

Steamjet cleaning was performed on all cornices, but was ineffective in the fifth floor cornice, just like carbonate products, due to the presence on the surface of a silicon-like film that had not been found in the sampled areas. It was not possible to complete the cleaning with other methods because the execution of the work was suspended for several months, pending the consensual termination of the contract with the contractor, which had entered into liquidation. The works then continued and finished by the company that ranked second in the public call for bids.

Figure 4.17 Lead plate with an inscription dated 24 September 1640, found on the roof in the southwest corner

The fourth-floor mullioned windows have large capitals made of granitello, a very compact and resistant variety of Aurisina stone, quarried in the area of the Adriatic coast near Triste and used since the 1st century BC by Romans, whose boats sailed upriver to reach the cities of the Po Valley. The capitals were affected by a very thick black crust whose concreted deposits were removed only minimally by steam cleaning, which, for this reason, was accompanied by the Jos system (a mixture of water and fine limestone powder with a low-pressure vortex to obtain a cleaning both gentle and easy to dose away from the surface).

The angular sculptures of the third cornice are partially made of granitello and partially of Vicenza stone, the latter being more degraded and having lost most relief.

The sculpture in the southeastern corner of the cornice, depicting a lion, had long been largely removed due to the danger of collapse. With the restoration it was possible to reassemble it by inserting long pins to anchor it to the wall. To better protect the most exposed sculptures and make them less prone to leaching, lead flashings that protrude from above the cornice were installed.

On the east side, three Vicenza stone–decorated panels of Roman origin were very degraded. They were first consolidated; then the black crusts were removed with laser technique. This methodology, both effective and respectful of the material, was only used for the most delicate and valuable sculptures because it requires long time for execution, specialized operators and has high costs (Fig. 4.18 and 4.19).

After cleaning, due to the widespread degradation of the surfaces, a very extensive plastering work was executed, with recipes tailored for each type of stone, to respect the colors and the compatibility with the material.

Figure 4.18 Detail of the head of the angular sculpture depicting a siren during the restoration with a laser technique

Figure 4.19 Detail of the same sculpture after the restoration with the laser technique

Note: The preservation of the ocher-colored oxalate patina can be observed.

Figure 4.20 First east-side frame, detail of red decoration depicting a flower and two lilies

Various cleaning techniques were used for cleaning a section of the first-floor cornice, on the east side, due to the presence of dripping from the cornice itself and black crusts more extended than on other floors. During washing with nebulized water, red hematite traces emerged. This was a decoration inside the arches, alternating a lily and a six-petal flower. The drawing, similar to others[9] dating back to the first half of the 13-th century, could be the work of the *Maestri Campionesi* (Fig. 4.20). Only three remain, but it is an important find because there was no previous evidence of external decoration of the tower in medieval times.

At the end of the work on decorations, both photographic survey and high-definition laser scan (Bertacchini *et al.*, 2015) were performed, making it possible to produce copies of the surveyed elements.

Because the conservation of the surfaces on the tower has been found to be better where the ancient protective patinas are still present, it was decided to apply a modern protective coat over the entire stone surface. The product is a microdispersion based on silanes and siloxanes, dissolved in water, that penetrates into the substrate, reducing the absorption of water on the surface but keeping the breathability unchanged. It is a sacrificial protective that, over time, will be washed away without causing any damage on the surface of the stone and has the purpose of making the results of the restoration last longer.

9 Similar red drawings were discovered inside Reggio Emilia Cathedral, on the outside of the church of San Michele in Foro in Lucca and in the church of Sant'Andrea in Vercelli.

4.4 Repairs and structural improvements

Geometric relief pointed out the tower's tendency to crack and open in the upper part, a phenomenon that could trigger a more serious failure in the event of an earthquake. A series of tests also allowed to investigate the anchors of the five orders of ties arranged in a sunburst pattern, belonging to the 16th-century elevation inside the spire, finding the presence of oxidation and fractures in the wall. The system, whose role is essentially to prevent the sliding of individual facades, was found to be not completely reliable due to these shortcomings.

At the top of the belfry is another tie system. Five of the ties are disposed on the same plane, but they are anchored to the internal walls of the angular pillars, which are hollow, without any overlapping and with lesions in correspondence with the anchors, making this system not completely effective.

The incomplete reliability of existing hoisting systems and the study on seismic vulnerability urged the need to create two new hooping systems complementary to existing ones. To avoid damage to the structure and to have reversibility that will allow possible new interventions in the future, it was decided to provide reinforcement with external hoops.[10]

The first hoop system is octagonal, was installed under a metal platform in the balcony at 60 m of height and is not visible. To accommodate the plates in the corners it was necessary to remove the connecting stones between the wall and the balcony and then to reassemble them.

The second, square-shaped stainless-steel system was placed at the base of the belfry. Plates 20-mm thick were used in the corners, isolated from contact with the stone by a layer of Teflon. Dimensions were determined by considering the maximum stress produced by expected seismic actions and by the wind (Figures 4.21 and 4.22).

The third to fifth floors are characterized by granitello (a type of Aurisina stone) slabs, with light-gray color and a typical granular appearance. The greatest number of hourglass injuries is found on these floors in the corners of the pilasters. To prevent further detachments, a simple operation was performed to attenuate the compressive stress peaks, which are at the origin of injuries and gaps: only where new fractures were triggered were cuts of a few centimeters of depth and a few millimeters of thickness made in correspondence of the horizontal joints, with a few centimeters of depth and a few millimeters of thickness cut. The cuts, 41 in total, were filled with lime and elastomer to better calibrate the stiffness and make it compatible with the structure also from a mechanical point of view.

The set of local repair measures, reinforcement especially, reduced some weaknesses and improved by no means the structure's behavior mainly in the event of an earthquake. The problem of the stability of the tower as a whole is addressed in Chapter 7, even if the issue of the interaction with the arches connecting the tower to the cathedral still deserve attention. This interaction acts in long-term movements of both structures, mainly considering the effects of subsidence (see Chapter 5) but may be more pronounced in seismic oscillation phase, as it was observed with the earthquake of Garfagnana in 2013 (Di Tommaso et al., 2013).

---

10 The hoop on the balcony and the lower hoop on the belfry cornice were designed by Prof. A. Di Tommaso and Prof. C. Blasi.

Figure 4.21 The tie rod at the base of the bells room

Figure 4.22 Detail of the angular plate

4.5 Bird deterrents

A specific project was planned with the Provincial Veterinary Service to reconcile the needs of restoration with maintaining of favorable conditions for the return of the colony of common swifts that nested in the tower before the scaffolding was installed.

At the same time, it was necessary to tackle the issue of a colony of stray pigeons because of the damage their droppings caused to decorated surfaces.

The numerous putlog holes on the façades are the traces of the wooden structures that were used to raise the buildings. Often these holes remained open even after the completion of construction works, and in the tower there are about 400, most of which have been used by birds as nesting sites. To favor swifts nesting over pigeons, the entrance was reduced in the upper floors' holes, according to a method already used in the past, which consists in inserting a wedge into the lower part of the hole by cutting the edge of a brick (Ferri *et al.*, 2015). This reduction from 10 cm to 4 cm allows access to the swifts but generally not to the pigeons. The remaining holes, positioned at the first levels, were closed with small steel nets.

An electric bird-deterring system was installed on all moldings and on most protruding sculptures. For the same reason, two-colored nylon nets to be barely visible from below were placed in front of the mullioned and single windows. Anti-bird copper nets that can be opened were placed over the large three-light openings of the bells room. One year after the completion of works, in the spring of 2013, the swifts started to visit the new reduced holes, while pigeons used other buildings. This attention during restoration contributed to protecting the urban biodiversity (Fig. 4.23).

During restoration works, thanks to the presence of a well-structured scaffolding, it was possible to organize numerous guided tours that gave the opportunity to more than 4,000 people to see the sculptures up close, to observe the techniques used for the restoration and to understand the need for long times of execution.

On 11 November 2011 the scaffolding was removed, and the tower was returned to the city with a celebration in the square.

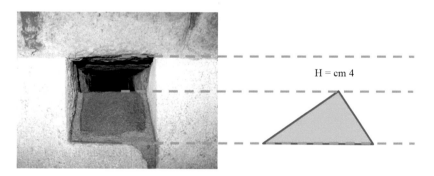

Figure 4.23 Example of the reduction of a putlog hole with the use of wedge bricks.

Note: The entrance hole is reduced from 10 cm to 4 cm in height.

4.6 Periodic checks

The Italian law on public works prescribes the drafting of a maintenance plan; however, this obligation is often only fulfilled as a generic formalism without any follow-up over time. If conservation of a monument starts from prevention, maintenance is a great way to deal with it.

In the case of an important monument such as the Ghirlandina Tower, a detailed plan based on the real control needs of a tall and unreachable tower was prepared. This is why, starting from the initial surveys, sample areas on which to carry out periodic checks and to compare the results over time were identified. The verifications are of different levels, from simple visual control to the instrumental analysis, precisely performed on sample areas.

Evaluation of surface protective systems, visual inspections of stone elements conservation state, verification of the cohesive state and the material continuity of stone elements, and evaluation of hygrometric content and superficial cohesive state of the plasters present at the inside of the cusp are periodically performed.

The results of these verifications provide a clear picture of how protective measures are decaying and, more generally, of the states of all interventions and are therefore particularly useful in conservation intervention planning, without waiting for new complete and expensive restorations.

4.7 Other interventions and new needs

The earthquake that severely damaged the towns north of Modena on 20 and 29 May 2012 also caused some damage to the tower. Surely the consolidation of the balconies, the two metal hoops and the vast plaster work contributed to reducing the effects of the earthquake on the tower.

Thanks to an intervention already planned, the vertical lesions that were already present on the south side on the fourth and fifth floors in correspondence with the openings were recomposed. The intervention consisted in inserting carbon strips in the horizontal joints between the bricks from the inside.

On the outside, the most evident damage is a new fracture, parallel to an already-restored one, on the sculpture, depicting Samson in the act of smashing the lion's jaw, placed in the southwest corner of the first cornice. The damage was caused by the hammering action between the tower and one of the arches connecting to the Duomo, which, once again, points out the interaction between the two monuments. Thanks to the comparison between the high-precision laser-scanner survey carried out at the end of the restoration and the one specifically repeated after the earthquake, it was possible to identify the type of movement in progress and to design the fixing operation that will take place in the next intervention. (Fig. 4.24 and 4.25).

In addition to a few easily legible lesions, however, there were many small cracks in the grouting and in all the elements in which fractures had been reduced.

For example, the disconnections along the joints of the stone elements of the balcony corner have very low sonic values that denote a deterioration of their stability. Even if stability is guaranteed by the presence of internal pins, the joints' grouting is crossed by cracks that are weak points and preferential ways for water infiltration that can activate the mechanical

Figure 4.24 Samson breaking the jaws of a lion, first floor, southwestern corner

Figure 4.25 Comparison between laser surveys performed before and after the 2012 earthquake

Source: Rivola *et al.* (2010).

Note: The fracture has triggered a wedge movement of escape from the plane (in red).

action of freeze–thaw in the winter months, initiating a disintegration process. For this reason, maintenance intervention to renew the grouting in the balconies and to keep connections efficient is necessary: the need for care never ends.

In 2015, thanks to a private loan, a good part of the tower guards' room was restored, including the exquisite capitals, one of the masterpieces of this tower. Also in this case, the removal of black crusts was performed with several techniques, including the use of rigid gels and laser.

Finally, in 2017 the entrance floor was restored with funding from a law for UNESCO sites and new interventions to complete the works and measures for securing the tower with increasingly accurate measures are at this time being planned.

Chapter 5

Soil as "material with memory"

A key to explain settlements of the tower and the cathedral

5.1 Factual data on the settlements of the tower

Investigations aimed at clarifying the founding depth and eventually the presence of piles, complemented with measurements of tilt, started at the end of the 19th century.

At that time (1898–1901) a pit was initially excavated near the southern side of the tower and the inspection revealed the *socle* of the ancient door at depth of 1.80 m from the ground surface (see the section sketched in Fig. 5.1).

Then, a trench was safely excavated at the northeast edge up to the depth of 4.90 m, where the *basolato* of the Roman road (*Via Emilia*) was found. By direct inspection, the presence of the *socle* at a depth of 1.36 m was observed, and the masonry of the foundation was noticed to reach a depth of 5.45 m. At that time the conclusion was reached about the absence of any piled foundation, but this was rather arbitrary because of a lack of any investigation beneath the founding level.

The Scientific Committee in 2007 realized the need for having a deeper knowledge of what the tower foundation is like, so new borings were planned, as sketched in Fig. 5.2.

Simply based on factual evidence, this investigation allowed to clarify many relevant aspects:

(a) the brickwork made foundation revealed to have a thickness of 3 m and was conceived as a spread foundation without supporting piles;
(b) the *socle* of the ancient door was found at a depth of 1.48 m from the actual ground level; and
(c) the boring G5 was intentionally drilled in such a way to intercept the *basolato* of the Roman road (*Via Emilia*) near the edge (at a depth of 5.45 m) and just below the foundation (at a depth of 6.75 m).

By comparing the depth of the *basolato* at 6.75 m with that of 4.90 m found during the investigation from 1898 to 1901, it could be argued that the tower suffered a settlement of the order of 1.85 m at the north side and, by considering the tilt, an average settlement slightly more pronounced.

This value represents a lower bound of the settlement the tower experienced because we can certainly observe that the depth of the *basolato* near the side of the foundation cannot be considered as representative of a *free-field condition*, that is the soil conditions not influenced by the presence of the tower. However, the boring E2, related to an archaeological investigation and far away from the tower, showed a depth of the *basolato* when referred to free-field

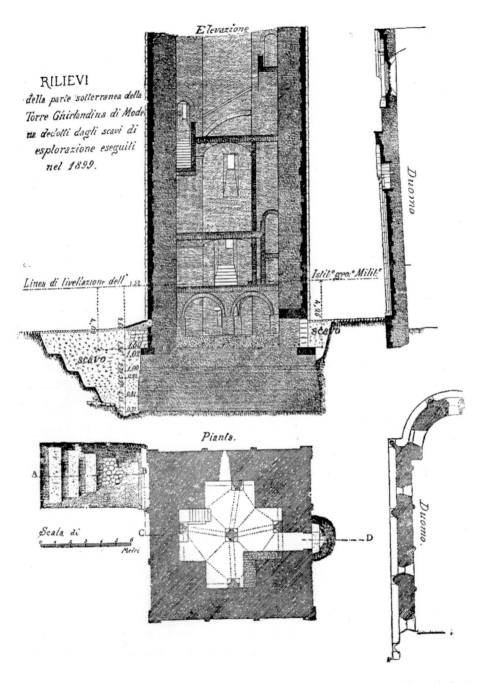

RILIEVI
della parte sotterranea della
Torre Ghirlandina di Modena dedotti dagli scavi di
esplorazione eseguiti
nel 1899.

Elevazione

Duomo

Linea di livellazione dell' 1.82

Istit.º geo.º Milit.

scavo

scavo

Pianta.

Scala di
Metri

Duomo.

Figure 5.1 Tower section as referred in the report on investigations performed during
1898–1901

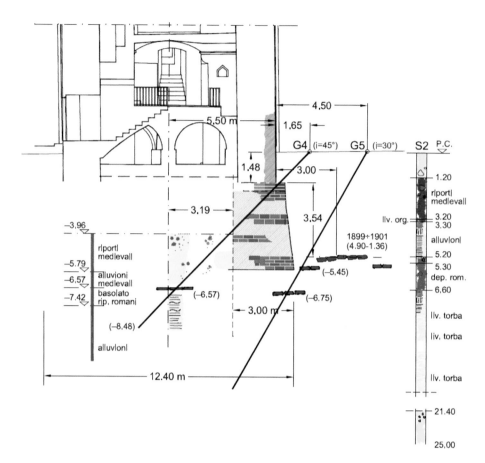

Figure 5.2 Soil profile and foundation geometry

Source: Lancellotta (2009a).

conditions equal to 4.80 m, and this result suggests an upper bound value of the average settlement of the order of 2 m.

This is certainly a rather pronounced settlement, whose entity can be explained by considering the nature of soil formation and the related geotechnical properties.

5.2 Site investigation

Since 1980, the municipality of Modena promoted studies related to the subsidence of the alluvial plain (Russo, 1985; Cancelli, 1986; Cancelli and Pellegrini, 1984; Pellegrini and Zavatti, 1980). In addition to these studies, the interest in archaeological sites suggested performing a lot of research on the Quaternary sedimentation of the Modena plain (Cremaschi and Gasperi, 1989; Fazzini and Gasperi, 1996; Lugli *et al.*, 2004). Finally, the need to assess the potential seismic vulnerability of the tower required a rather comprehensive site investigation, which was planned in September 2007 and December 2008 as described in detail by

Lancellotta (2009a). Therefore, at present we are able to summarize the relevant geotechnical aspects as follows.

By referring to Figure 5.3, the soil profile down to the investigated depth of 80 m is composed of a first horizon of medium- to high-plasticity inorganic clays, with an abundance of laminae of sands and peat, only millimeters thick. The upper portion of this horizon, whose

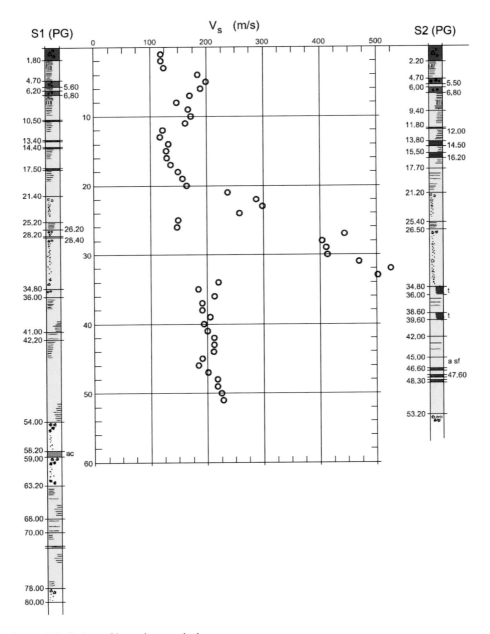

Figure 5.3 Soil profile and cross-hole tests

Source: Lancellotta (2009a).

thickness ranges from 5 m to 7 m, is known as the Modena Unit and is linked to the flooding events (of post-Roman age) produced by minor streams (Fossa-Cerca).

The subsequent underlying horizons, ranging in depth from 21.40 m to 54 m, represent the result of a complete transgressive-regressive cycle: the fine-grained sediments, belonging to the horizon known as Niviano Unit, were deposited during the penultimate interglacial cycle, and the superimposed coarse-grained materials, belonging to the Vignola Unit, are linked to transport activities of the Secchia River.

A second horizon of coarse-grained materials is encountered at depths ranging from 54 m to 63.20 m, and thereafter, a fine-grained horizon is found down to a depth of 78 m, here again characterized by a diffuse presence of laminae of sand.

The soil profile, shown in Figure 5.3, has been complemented with the results of cross-hole tests.

These tests are performed by producing seismic energy in a borehole and the time the seismic wave takes to reach another borehole is measured (Fig. 5.4.a).

Two boreholes are usually sufficient to perform the test, but to eliminate the errors due to the triggering of the timing instruments, a preferable array is represented by three or more boreholes so that the wave velocity Vs can be computed from the time intervals between pairs of holes.

The arrangement for the *down-hole test* is also shown in Figure 5.4.b. In this case, the impulse is generated at the surface and the receivers are clamped to the borehole at different depths. This offers the advantage of being a more economical test arrangement, if compared to cross-hole tests, but care is required with data interpretation.

Because of their noninvasive character, these tests allow the preservation the initial structure of soil deposits as well as the influence of all diagenetic phenomena (sutured contacts of grains, overgrowth of quartz grains, precipitation of calcite cement and authigenesis) contributing to a stiffer mechanical response. For this reason, they represent the most reliable methods of determining the shear modulus at small strain amplitude.

Because the strain level is less than 10^{-3}%, a reasonable assumption is to analyze the results of these tests by referring to the wave propagation theory in elastic materials. Accordingly,

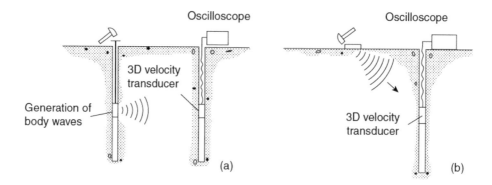

Figure 5.4 Cross-hole and down-hole tests

Source: Lancellotta (2009b).

the soil stiffness G, indicated as G_0 when referred to small strain level, is inferred from measurements of shear wave velocity, by using the relation

$$G_o = \rho \cdot V_s^2 , \qquad (5.1)$$

where ρ is the soil density and V_s is the measured shear wave velocity.

More insight into explaining the huge settlement suffered by the tower and the differential settlements of the cathedral requires a short digression on soil mechanics and, in particular, on soil behavior intended as the heritage of the history of previous events (see for more details Lancellotta [2009b] or other textbooks on soil mechanics).

5.3 A digression on soil mechanics: the soil as "material with memory"

Soils are composed by an aggregate of solid particles with voids or pores between particles that are filled with water if, for sake of simplicity, we refer to fully saturated soil, as it is for the present case. It is rather intuitive to say that mechanical properties of engineering relevance, such as shear strength and compressibility, depend on the degree of packing of the particles so that it is of interest to define in some way the amount of packing. Particles can be considered as incompressible in geotechnical applications, and volume changes are produced by a reduction of the volume of voids; therefore, it is usual to assume as a state variable, that is, a variable that indicates the current state of the soil in terms of aggregation of particles, the *void ratio* (we use the lowercase e to indicate this variable), defined as the ratio between the volume occupied by voids V_v and the volume occupied by the solid particles V_s:

$$e = V_v / V_s . \qquad (5.2)$$

When soil is loaded for the first time to a stress level greater than it has previously experienced, the void ratio progressively reduces with the applied vertical effective stress σ_v' (path a and b in Fig. 5.5).

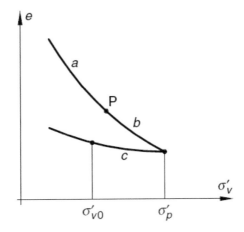

Figure 5.5 Stress history of a soil sample

Such a soil is defined to be *normally consolidated* and the curve joining successive point representing such states is termed *virgin compression* or *consolidation line*.

If the soil element is unloaded, as it is the case with natural processes, such an erosion of overlying sediments, or with a man-made process, such an excavation, it increases in volume along the swelling line (path *c* in Fig. 5.5) but at a much slower rate. A soil represented by a current state belonging to this unloading or swelling line is said to be *overconsolidated*. The maximum vertical effective stress σ_p', which acted over its geological and the most recent stress history, is defined as *preconsolidation stress*. Furthermore, in order to quantify the entity of the overconsolidation process, the *overconsolidation ratio* (*OCR*) is introduced as a convenient parameter, defined by the ratio between the maximum past stress and the actual overburden vertical effective stress:

$$OCR = \sigma_p' / \sigma_{vo}'. \tag{5.3}$$

The stress history of an undisturbed soil sample is usually investigated in the laboratory by performing a compression test in a special testing apparatus called *oedometer* and Fig. 5.6 shows a typical result in terms of current void ratio *e* plotted *versus* the logarithm of the applied vertical effective stress, as it is common in soil mechanics.

If we consider the compression curve within the context of the plasticity theory, we can attribute to the break point *B* to the meaning of a *yield stress* because it separates the small-strain behavior from large strains.

Accordingly, the first portion of the compression curve, from point *A* to *B*, is defined **recompression curve**, and soil behavior is assumed to be nonlinear but almost completely reversible along this path. The second portion, from *B* to *C*, is called **compression curve** and

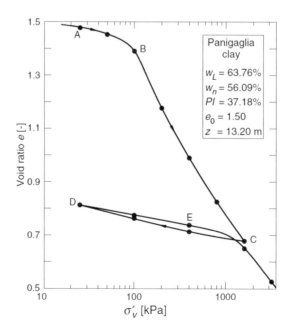

Figure 5.6 Typical compression curve on an undisturbed clay sample

Source: Lancellotta (2009b).

is characterized by largely irreversible strains, as can be proved by unloading the specimen from C to D.

The soil has *memory* of its previous loading history, as it is revealed from the fact that, when the specimen is further reloaded starting from D, its behavior appears to be reversible until the current yield stress has been attained.

The described behavior has two important implication on engineering practice:

(a) first, when the soil will be subjected to a reloading process, deformations will be elastic until the applied vertical effective stress will not exceed σ'_p, whereas for greater loads plastic deformations will play the major role in addition to the elastic ones, and

(b) as a corollary, it follows that the assessment of the preconsolidation (or yield) stress is the single most important step when characterizing the soil behavior, and it is of relevance to recall that the trend of yield stress and overburden effective stress with depth represents a concise picture of the stress history of the deposit (Jamiolkowski *et al.*, 1985; Lancellotta, 2009b).

For example, a *mechanical overconsolidation* (i.e. removal of previous loads or erosion) is characterized by a constant difference $\left(\sigma'_p - \sigma'_{vo}\right)$ with depth; if the preconsolidation is ascribed to aging, a constant ratio σ'_p / σ'_{vo} with depth should be expected; a desiccation due to drying and freeze–thaw cycles produce scattered values of σ'_p; physio-chemical phenomena can produce an increase of yield stress, with a quite variable profile.

Different mechanisms can obviously act together, and quite often the stress history profiles are not so simple. This is the case under consideration because when considering the trend

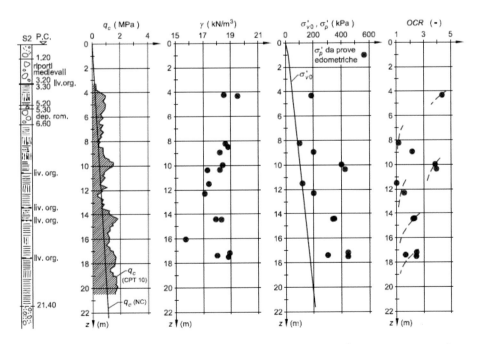

Figure 5.7 Stress history of Modena clay Overburden stress (σ'_{vo}), yield stress (σ'_p) and overconsolidation ratio (OCR) γ = unit weight; q_c = cone resistance

Source: Lancellotta (2009a).

of yield stress and overburden effective stress with depth shown in Figure 5.7, as well as the trend of the overconsolidation ratio OCR on the same figure, no simple arguments are able to justify this apparently erratic trend.

More insight in this case was gained by considering on the same figure the trend of cone penetration resistance. Briefly, the cone penetration test consists of pushing a conical point (see Fig. 5.8) at a constant rate of 20 mm s^{-1} and the tip resistance (which is indicated in Fig.

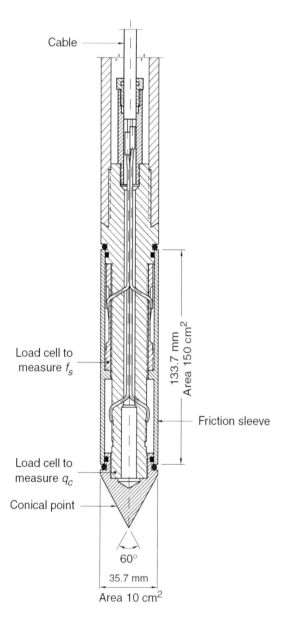

Figure 5.8 Typical design of electrical cone penetrometer

5.7 with the symbol q_c) is measured. This test is characterized by high repeatability and high accuracy, and the possibility of including additional sensors, such as those to measure the pore water pressure during the penetration, gives this test the capability of detecting small lenses and macrofabric features in soil profiling.

In normally consolidated soil, the cone resistance increases linearly with depth; therefore, it can be presumed that any deviation from this trend is certainly a matter of overconsolidation phenomena. In addition, by observing in Figure 5.7 the cyclic character of such deviations, it can be argued that the presence of preconsolidated soil layers can be related to exposure and desiccation during the deposition cycles, as also confirmed by a diffuse pedogenesis observed on soil samples of high quality.

Based on the above consideration that soil has "memory" of previous stress history, we now come back on the profile of settlements of the cathedral and the tower, by taking into account the aspects related to the construction of the cathedral, discussed in Chapter 2.

In particular, the process of building and dismantling the previous cathedral (see Fig. 2.4) induced on the supporting soil a loading and unloading process in the western zone so that part of the Lanfranco Cathedral was built on less compressible soils. On the contrary, the apses were built on a virgin, more compressible soil. This explains the rotation of the apses of the cathedral toward the east (Fig. 5.9) and not only toward the north, as a result of the interaction with the Ghirlandina Tower (Fig. 5.10).

Furthermore, by considering the compression tests performed on undisturbed samples of Modena clay (see Fig. 5.11 and Tables 5.1, 5.2) and taking into account that the contact stress applied by the tower foundation is equal to 554 kPa (corresponding to a dead weight equal to 85.24 MN), it can be explained the pronounced settlement experienced by the tower.

5.4 Further insights into the settlements of the tower: the subsidence phenomenon

In addition to the previously mentioned total and differential settlement experienced by the cathedral and the tower during the construction stages and the subsequent decades, in more recent years additional settlements were induced by the subsidence phenomenon.

To illustrate this phenomenon, Figure 5.12 considers a clay stratum underlaid by a sand aquifer. If the groundwater level in the upper sand horizon is constant due to recharge, whereas in the underlying aquifer is significantly lowered below that in the upper one, for water supply purposes, when the steady-state condition will be reached, the pore water pressure (indicated by the lowercase letter u) distribution within the clay stratum will be that shown by the line AB. The reduction of the water pressure with respect to its initial value (represented by the line AC) will cause an equivalent increase of effective stresses, which will give rise to a compression of the underdrained soil and subsidence of its surface.

Studies based on topographic leveling carried out by the Italian Military Geographical Institute (IGM) between 1887 and 1889 and 1949 to 1950, a period in which it can be supposed that the Modena area was not interested in subsidence induced by anthropogenic causes, suggest a natural subsidence rate of the order of about 2.5 mm y^{-1}. Subsequently, in 1981 the municipality of Modena promoted a comprehensive study of the induced subsidence that resulted in the creation of a local, dense network of geodetic leveling (Russo, 1985) and in the monitoring of the piezometric levels of both the surface groundwater and the groundwater related to deeper aquifers. These studies have shown that the deep-aquifer

Figure 5.9 Differential settlements and tilt toward the east due to the pre-loading imposed by previous cathedrals

level, which originally exceeded the ground level, is subsiding over time because of the strong water withdrawals, resulting in the late 1970s to about 10 m below ground level in the most industrialized areas. The increase of subsidence induced by this withdrawal reached peaks ranging from 60 cm in the historic center of Modena to about 80 cm at the northern part of it (Fig. 5.13).

Figure 5.10 The tilt of the apses toward the Ghirlandina Tower

In the following years, thanks to the initiatives of the municipality of Modena to reduce water withdrawals, the piezometric level of the aquifers located between 22 m and 34 m as well as the ones between 54 m and 63 m depth, that in 1975 and 1976 stood at 10 m from the ground surface, rose to 3 m to 4 m in the late 1980s. Today, recent measurements performed

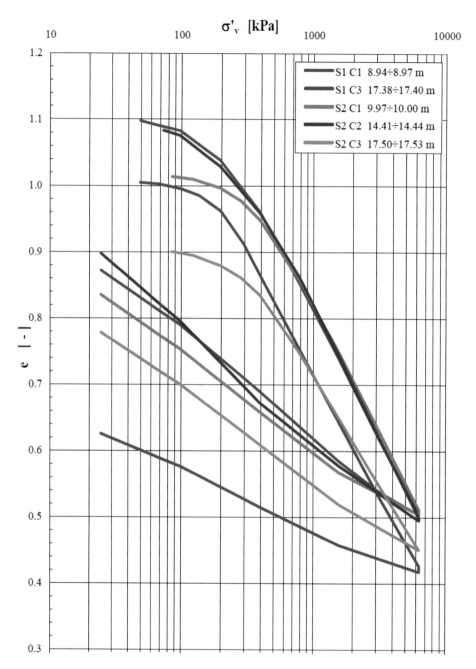

Figure 5.11 Compression tests on Modena clay
Source: Lancellotta (2009a).

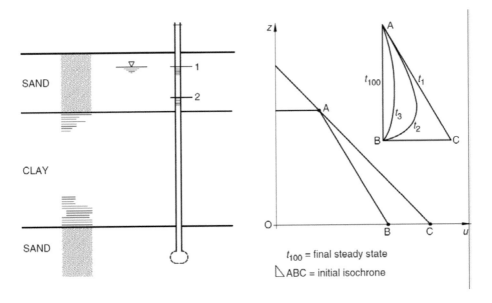

Figure 5.12 Illustrative example of subsidence induced by underdrainage

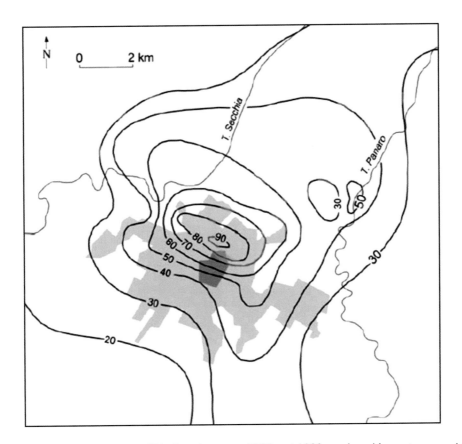

Figure 5.13 The subsidence of Modena between 1950 and 1980 produced by water pumping

Source: Castagnetti et al. (2017).

Note: Contours are reported in centimeters; courtesy of internal archive of Modena municipality.

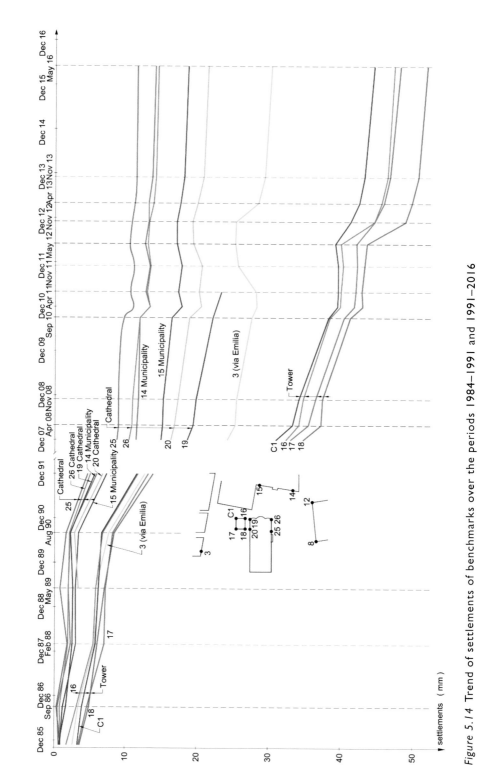

Figure 5.14 Trend of settlements of benchmarks over the periods 1984–1991 and 1991–2016

Source: Castagnetti *et al.* (2017).

in the historic center place this level to about +0.5 m from the ground level, whereas within the first horizon the phreatic groundwater level oscillates from a depth of 1.5 m to 1.2 m.

The geodetic leveling network consists of about 280 benchmarks and is connected to the national infrastructure. Furthermore, a densification was established in 1984 and surveyed until 1991 in order to control the settlements in the historic center nearby the tower and the cathedral.

Unfortunately, there was a period of about 15 years of interruption of such leveling, but due to the impact of the phenomenon on the cultural heritage, the municipality of Modena decided in 2007 to restart the high-precision leveling, and the available results are shown in Figure 5.14.

It is apparent that the tower suffers settlements more pronounced than the *Piazza Grande* square and the cathedral as well shows significant differential settlements between the south and north apses.

Table 5.1 Index properties, stress history and soil compressibility of Modena clay (soil investigation performed in 2009)

Bore-hole	Sam-ple	z (m)	γ (kN/m³)	G_s (-)	w (%)	w_L (%)	PI (%)	CF (%)	CaCO₃ (%)	σ'_{vo} (kPa)	e_o (-)	σ'_p (kPa)	OCR (-)	C_c/C_s (-)
S2/09	C1	4.32	19.5	2.74	27.3	35.0	12.7	26.0	22.9	49.31	0.755	185	4.36	0.24 / 0.033
S3/09	C3	8.32	18.6	2.73	34.1	40.8	18.4	25.3	–	86.93	0.921	100	1.15	0.256 / 0.039
S2/09	C3	10.36	18.2	2.73	37.4	73.9	40.2	61.3	13.6	108.5	1.019	426	3.93	0.422 / 0.11
S1/09	C2	11.50	17.4	2.75	46.5	76.8	43.6	65.6	14.0	120.5	1.267	120	1	0.422 / 0.115
S2/09	C4	12.31	17.1	2.74	48.9	67.2	37.1	51.8	–	129.1	1.333	200	1.55	0.57 / 0.047
S1/09	C3	16.04	15.7	2.72	63.5	79.8	44.8	69.6	12.1	168.5	1.782	–	–	0.589 / 0.144

Source: Lancellotta, 2009.a.

Note:
z: depth below ground level
γ, G_s: unit bulk weight and specific gravity of soils
w_N, w_L: natural water content and liquid limit
PI: plasticity index
CaCO₃: calcium carbonate content
e_o: initial void ratio

σ'_p, C_c, C_s: yield stress and compressibility index (C_c: compression index; C_s: swelling index)

CF: clay fraction

Table 5.2 Index properties, stress history and soil compressibility of Modena clay (soil investigation performed in 2007)

Bore-hole	sam-ple	z (m)	γ (kN/m³)	G_s (-)	w_N (%)	w_L (%)	PI (%)	CaCO₃ (%)	σ'_{vo} (kPa)	e_o (-)	σ'_p (kPa)	OCR (-)	C_c (-)
S1/07	C1	8.94	18.2	2.74	36.3	57.8	28.9	20	93.48	1.01	200	2.14	0.37
S2/07	C1	9.97	18.4	2.77	36.3	101.5	61.3	7.7	104.4	1.01	400	3.83	0.38
S2/07	C2	14.41	17.9	2.72	40.4	89.5	54.8	25.3	151.3	1.09	347	2.29	0.40
S2/07	C2	14.44	18.3	2.72	39.1	89.5	54.8	–	151.6	1.03	333	2.20	0.39
S1/C3	C3	17.17	18.9	2.77	32.5	85.1	52.1	–	182.6	0.90	447	2.45	0.34
S1/07	C3	17.38	18.0	2.77	39.3	85.1	52.1	22.4	182.6	1.09	300	1.64	0.39
S2/07	C3	17.50	18.8	2.74	32.9	94.2	59.1	12.3	183.9	0.90	447	2.43	0.33

Source: Lancellotta, 2009.a.

In particular, by considering the 1985–2016 interval, the average rate of settlement of the Ghirlandina Tower was about 1.5 mm y^{-1}, but in more recent years (interval 2013–2016) a reduced rate of about 1.19 mm y^{-1} is being observed.

These values are well below those reached during the 1970–1980 interval, when the subsidence reached its peak, and prove the effectiveness of the initiatives of the municipality of Modena to reduce water withdrawals, the difference being even more pronounced in terms of rate of tilting of the tower.

These conclusions show how important can be to constantly and carefully monitor the two structures and the hydraulic boundary conditions through piezometric measurements in order not only to verify our predictive models (numerical or simply conceptual) but also to check the effectiveness of protective measures.

Chapter 6

A primer on soil–structure interaction and soil condition effects

6.1 Introductory notes

As already outlined in Chapter 1, there has always been a focused interest in the potential seismic vulnerability of historic towers. In the case of Ghirlandina Tower, this interest was further increased by the recent earthquake events in the Emilia Romagna region in May 2012 because many historical masonry *campaniles* (bell towers) collapsed despite the moderate magnitude (M = 5.9) of these events.

To properly capture the behavior of a structure, particularly of a historic tower, during seismic events, it is of paramount importance to take into account the interaction with the supporting soil and the effects related to soil conditions, that is, how soil properties may influence the input motion at the base of the tower.

The reader who is unfamiliar with concepts of dynamics may prove to be difficult to grasp these aspects, even in a rather simplified version, and having in mind these difficulties, we considered beneficial to have in this chapter a remind of basic aspects and to provide a lexicon that can help to appreciate the more specialized topic addressed in Chapter 7.

6.2 The single degree of freedom system

Consider the single degree of freedom (SDOF) system represented in Figure 6.1, where a mass m is attached to a massless column of height h. The column is flexible to horizontal deflexion and is characterized by a horizontal *stiffness* k, and we assume at this stage that its base is fixed on firm ground; that is, it is restrained against any translation and rotation.

This idealized system may be regarded as a simple one-story building or as the model of a multimode structure which oscillates in its fundamental natural mode. In this latter case, the height h has to be considered as the distance of the centroid of the inertial forces, and m is the associated generalized mass.

If the mass is initially displaced to some lateral distance u_o, it will start to oscillate, and experience shows that the amplitude of these oscillations will be ever decreasing, and eventually the mass will come to rest.

To take into account of this behavior, an energy-absorbing element needs to be introduced, and for mathematical convenience, this is usually represented through a *viscous damper* so that the same system can be represented as in Figure 6.2.

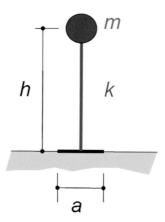

Figure 6.1 The SDOF system

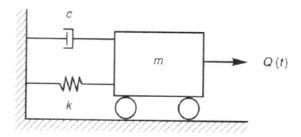

Figure 6.2 The equivalent SDOF system

Fig. 6.3 shows the forces acting on the system when it is pushed away from its initial equilibrium position by a time-dependent force $Q(t)$:

(a) the elastic force due to the stiffness of the spring is given by $f_s = k \cdot u(t)$,
(b) the inertial force is proportional to the acceleration $\ddot{u}(t)$ and is given by $f_i = m \cdot \ddot{u}(t)$ and
(c) the viscous damper acts with a force proportional to the velocity $\dot{u}(t)$ and is given by $f_D = c \cdot \dot{u}(t)$.

Then, the equation describing the deformation of the idealized structure can be written as

$$m \cdot \ddot{u}(t) + c \cdot \dot{u}(t) + k \cdot u(t) = Q(t). \tag{6.1}$$

To examine the more complex case of an earthquake-induced vibration at the base of the structure in Figure 6.1, suppose that $u_g(t)$, $\dot{u}_g(t)$, $\ddot{u}_g(t)$ are the horizontal component of the displacement, the velocity and the acceleration of the ground motion.

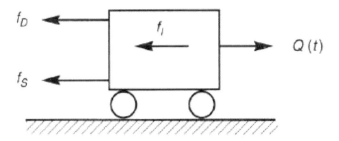

Figure 6.3 Forces acting on the system of Figure 6.2

In textbooks on the dynamics of structures (see Clough and Penzien, 1993; Chopra, 1995), it is proved that the Equation (6.1) still applies, provided that the external force $Q(t)$ is substituted by an effective force equal to $-m \cdot \ddot{u}_g (t)$:

$$m \cdot \ddot{u}(t) + c \cdot \dot{u}(t) + k \cdot u(t) = -m \cdot \ddot{u}_g (t). \qquad (6.2)$$

Consider now the idealized structure without any damping. If the mass is moved away from its initial position of equilibrium, the motion does not decay, and the displacement amplitude is the same for all cycles of vibration. Then, the time T (seconds) required for one cycle of free vibration is called the *natural period of vibration*, and it is related to the *natural circular frequency* of vibration ω (radians/seconds) or to the *frequency of vibration* f (cycles per seconds or Hz) by the relation

$$T = \frac{2\pi}{\omega} = \frac{1}{f}. \qquad (6.3)$$

Note that these quantities are properties of the system when it is allowed to vibrate freely without external excitation and this explains the use of the adjective *natural*. Because the system is supposed to be linear, these properties are independent on initial displacement or velocity, and therefore, it can be proved that the previously defined quantities only depends on the mass and the stiffness of the system; that is,

$$\omega = \sqrt{k / m}. \qquad (6.4)$$

In contrast to this idealized system, the more realistic damped model oscillates with amplitudes decreasing with every cycle of vibration, and in this case, the natural circular frequency of vibration ω_D will be influenced by the damping according to the following relation:

$$\omega_D = \omega \cdot \sqrt{1 - \xi^2}, \qquad (6.5)$$

where

$$\xi = c / (2m\omega) \qquad (6.6)$$

is a dimensionless measure of the damping of the system and is called *damping ratio*.

Note that for damping ratio lesser that 0.2 the differences between ω_D and ω are negligible, so it is common to assume

$$\omega_D \cong \omega \text{ and } T_D \cong T.$$

6.3 The response spectrum

The solution of Equation (6.2) provides the time history of the displacement response $u(t)$, which depends on the natural period of the system $T = \dfrac{2\pi}{\omega} = \dfrac{1}{f}$, the damping ratio $\xi = c/(2m\omega)$ and the ground acceleration $\ddot{u}_g(t)$.

However, in current design practice, it is sufficient the knowledge of the maximum value of the response, given a time history of the seismic event.

The plot of this maximum value (without regard to algebraic sign) as a function of the natural period (or of the natural frequency) of the structure is known as *response spectrum*.

Therefore, the *displacement response spectrum* is a plot of the quantity

$$S_d = u_{max} \tag{6.7}$$

obtained by solving the Equation (6.2) for a range of values of the period T (or the natural circular frequency ω), while keeping constant the damping ratio ξ.

The maximum value of the base shear force is therefore

$$V_{max} = k \cdot S_d, \tag{6.8}$$

and by taking into account the Equation (6.4), it can also be written as

$$V_{max} = k \cdot S_d = m \cdot \omega^2 \cdot S_d = m \cdot S_a, \tag{6.9}$$

where the introduced quantity

$$S_a = \omega^2 \cdot S_d \tag{6.10}$$

has the unit of acceleration and is called *pseudo-acceleration*, the prefix *pseudo* being used to distinguish this quantity from the true or correct acceleration (see page 228 of Chopra, 1995). The two definitions differ from each other by the term $2\xi\omega \cdot \dot{u}$ and are coincident if the damping ratio vanishes.

Similarly, the maximum response can also be expressed in terms of the so-called *pseudo-velocity*, defined as

$$S_v = \omega \cdot S_d, \tag{6.11}$$

a quantity related to the maximum strain energy E_{max} stored in the structure during the earthquake:

$$E_{max} = \frac{1}{2}k \cdot S_d^2 = \frac{1}{2}mS_v^2. \tag{6.12}$$

An illustrative example of response spectra of such quantities is given in Figure 6.4, and it must be outlined that any one of these spectra can be obtained from one of the other two and contains the same information.

We spent this introductory remind because the response spectrum approach is, at present, widely used for the design of earthquake-resistant structures, and although more advanced design tools are advocated for performance-based design of structures experiencing a nonlinear response, it remains a central concept in earthquake engineering design.

A detailed history about the origin and subsequent developments of the response spectrum design method is given by Trifunac (2006a, 2006b), and from this history a peculiar aspect emerges that has to be mentioned.

The contributions of M.A. Biot on many fields of geotechnical engineering are widely recognized (these include well-known papers on the theory of consolidation and wave propagation in porous-elastic media), but rather surprisingly, it is less recognized that M.A. Biot also made a pioneering contribution to the concept of the response spectrum in the early 1930s.

The concept of how to estimate the maximum response of oscillators to a transient excitation was in fact first described by Biot in Chapter 2 of his PhD thesis, developed under the supervision of Professor Theodore von Kármán at the California Institute of Technology.

At that time and further in early 1940s, Biot was able to describe the computation of response spectra by means of a mechanical analyzer and to formalize the general theory of response spectrum by emphasizing the use of a standard spectral curve, obtained as the envelope of a number of seismogram spectra, for the evaluation of the maximum effect on buildings (Biot, 1941, 1942).

Some years later Housner (1959, 1970) noted the response spectra of four strong motion records (El Centro, 1934; M = 6.5; El Centro, 1940, M = 6.7; Tehachapi, 1952, M = 7.7; Olympia, 1949; M = 7.1) and suggested the use of an averaged spectrum in engineering design.

In the early 1970s, the dependence of spectral shapes on the local site conditions was clearly put into evidence by Hayashi *et al.* (1971). They suggested three groups of averaged spectra from 61 accelerograms, depending on site conditions, and their results were further corroborated by Seed *et al.* (1976) on the basis of 104 records.

6.4 The relevance of soil–structure interaction

Moving along this path, in a pioneering work published in 1954 Housner was outlining the following:

> *Regarding the effect of the soil on the natural period of vibration, it is easily seen that, if the soil is sufficiently firm, a structure resting on it may be considered rigidly supported at its base. However, a soft soil may permit deflections of the footing and, thus, allow the structure to rotate about its base. This will increase the period of vibration of the structure which, in general, has a beneficial effect upon the stresses produced during an earthquake.*

Since that time, it is widely recognized that the response of a structure resting on a soft soil may be significantly different from the response of an identical structure supported on firm soil.

As outlined by Veletsos and Meek (1974) there are two major factors that explain this difference. First, the structure on soft soil has more degree of freedom and therefore different

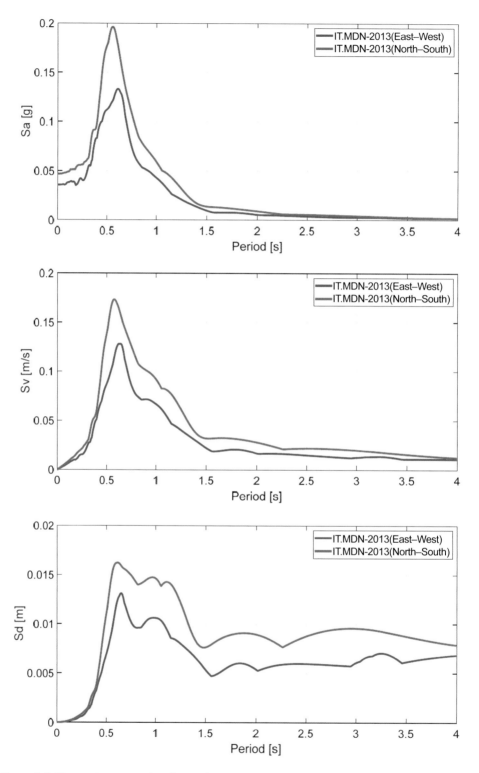

Figure 6.4 Illustrative example of pseudo-acceleration, pseudo-velocity and displacement spectra for the seismic event of 21 June 2013 in Modena

dynamic characteristics. Second, most of the vibration energy is being dissipated into the supporting soil by radiation and by internal damping of the soil itself.

To take into account the interaction of the structure with the supporting soil, a simplified but effective approach considers the soil response as lumped into a series of springs and dashpots, each of them related to the degree of freedom of the soil-foundation system.

It is important to note that the properties of these elements, springs and dashpots, depend not only on the properties of the half-space representing the soil but also on the exciting frequency.

By relying on this approach, if the soil and the foundation are represented by a horizontal spring and a rocking spring of stiffness K_h and K_r, $T_o = 1/f_o$ is the natural period of the structure resting on a fixed basis and ξ is its damping ratio, the interaction with the soil will give rise to a modified period of vibration T^*, given by

$$T^* = T_o \sqrt{1 + \frac{K}{K_h} + \frac{Kh^2}{K_r}}. \qquad (6.12)$$

Similarly, the equivalent damping ratio of the SDOF representing the structure will be given by

$$\xi^* = \xi_{soil} + \left(\frac{T_o}{T^*}\right)^3 \xi, \qquad (6.13)$$

where the contribution of the soil damping ξ_{soil} includes both the radiation and the material damping.

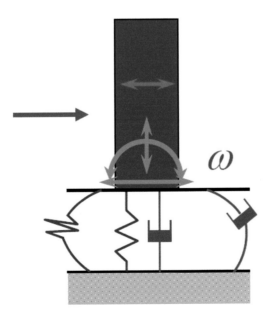

Figure 6.5 Simplified model of soil–structure interaction

6.5 A mathematical representation of soil response

When considering a harmonic excitation characterized by a circular frequency $\tilde{\omega}$, the ratio between the steady-state force (or moment) and the displacement (or rotation) of a massless foundation is called *dynamic stiffness*.

The dynamic force and the displacement are generally out of phase, and the dynamic displacement can be split into two components: one in phase and one 90 degrees out of phase with the imposed harmonic force.

Therefore, it is a matter of mathematical convenience to represent the dynamic response of the soil-foundation system in the form

$$\tilde{K}_j = K_j^s(k_j + i\,a_o c_j), \tag{6.14}$$

with K_j^s being the static stiffness of the foundation-soil system; a_o is a dimensionless frequency parameter depending on the exciting frequency $\tilde{\omega}$, the radius R of the foundation and soil properties.

$$a_o = \frac{\tilde{\omega}R}{V_s}; \tag{6.15}$$

k_j, c_j are frequency-dependent coefficients; and i is the *imaginary unit* with the property $i^2 = -1$.

The supporting soil is characterized by means of the shear wave velocity V_S.

The real component in Equation (6.14) reflects the stiffness and inertia of the soil, and its dependence on frequency comes from the effects that frequency has on the inertia because the soil properties are frequency independent.

The imaginary component represents the radiation damping and the material damping. The radiation component represents the energy dissipation by waves propagation away from the foundation and is frequency-dependent. The material damping reflects the hysteretic cyclic behavior of the soil and is frequency-independent.

By considering the analogy between the SDOF system and the massless foundation, the dynamic stiffness can also be represented in the form

$$\tilde{K}_j = K_j^{real} + i\tilde{\omega}C_j, \tag{6.16}$$

where K_j^{real} is the stiffness of an equivalent spring and C_j is the constant of an equivalent viscous dashpot.

Note that the dynamic coefficient c_j (with a lowercase letter) in Equation (6.14) should not be confused with the dashpot constant C_j (with a capital letter) in Equation (6.16), and the following relationship applies:

$$C_j = \frac{K_j^s \cdot R}{V_s} \cdot c_j. \tag{6.17}$$

As an example, if we consider a circular footing of radius R, resting on a half-space characterized by a shear modulus G and a Poisson ratio υ, the static stiffness for horizontal, vertical and rocking motion assume the following expressions:

$$K_h^s = \frac{8GR}{2-\upsilon} \qquad K_v^s = \frac{4GR}{1-\upsilon} \qquad K_r^s = \frac{8GR^3}{3(1-\upsilon)} \tag{6.18}$$

and the frequency-dependent coefficients to be introduced in Equation (6.14) are plotted in Figure 6.6.

Note that if the soil layer has a finite depth and rests on a much stiffer material, for example a bedrock like material, (a) there will be an increase in the values of the static stiffness; (b) there will be a stronger dependence of the dynamic coefficients on frequency, due to the existence of natural frequencies of the soil deposit; and (c) the damping component will be zero below the appropriate frequencies for each mode (see Gazetas, 1983, 1991).

By summarizing this paragraph, it appears, by considering both Equations (6.12) and (6.13), that the effect of the interaction is to increase the resonant period of the structure, with the benefit that the spectral acceleration to be used in design will be reduced (see Fig. 6.7); furthermore, the introduction into the analysis of the material damping of the half-space and the effect of energy radiation will result in an increase of the effective damping of the interacting system and a decrease of its response.

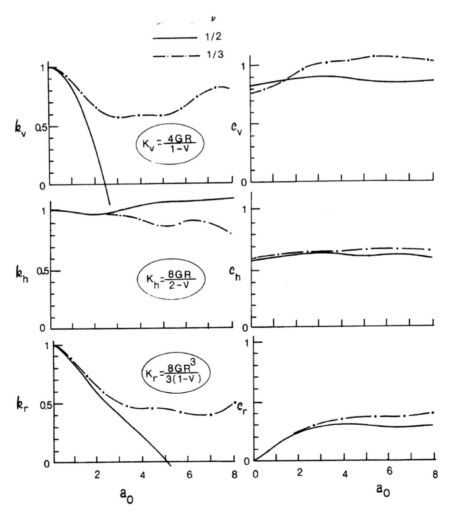

Figure 6.6 Frequency-dependent coefficients (Equation 6.14) of rigid circular foundation on homogeneous half-space (adapted from Gazetas, 1983)

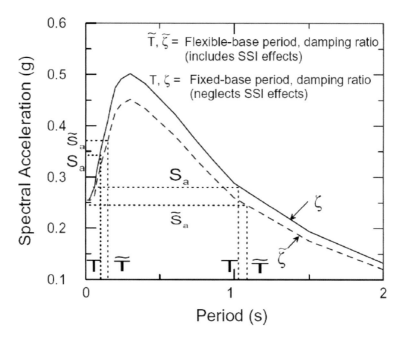

Figure 6.7 Influence of soil–structure interaction

6.6 Effects of local soil condition on earthquake-induced motion

All the considerations we made in Section 6.4 refer to what is called in literature *inertial interaction* because the soil–structure interaction is caused by the inertial forces generated by the vibrations of the structure. These forces produce base shears, axial forces and moments acting on the foundation, and the resulting soil deformations will alter again the motion of the foundation.

However, there are still two other effects to be recognized.

If we consider a massless foundation, before the structure is built, or may be a structure with no mass, the motion of the foundation will be different from the motion recorded at the free surface of the soil and the interaction responsible of this difference is called *kinematic interaction*.

Rigid surface foundations, subjected to a train of plane waves propagating at arbitrary angles of incidence will experience a reduction of the translational motion in the high-frequency range and in addition the appearance of torsional components of motion, that is, a rotation around a vertical axis. Only when dealing with surface foundations and vertically propagating waves the two motions of the free field and the foundation will be equal.

If we are dealing with embedded foundation, the kinematic interaction will play a role also for vertically propagating waves. More generally, there will be a filtering effect of the translational motions at high frequencies and the appearance of the rotational component of the motion, that is, a rocking around a horizontal axis.

Finally, earthquakes recorded at different sites show very different characteristics, in particular, frequency content, as it was apparent with accelerograms recorded in the valley of Mexico City.

This aspect, mentioned in literature as *local seismic response* or *soil amplification*, was recognized in 1957 by Kanai, who first suggested a simplified model for soil amplification studies, and this approach was complemented in the late 1960s by many researchers.

These studies considered a horizontally stratified soil deposit with properties varying only in the vertical direction and shear waves propagating vertically. Therefore, these assumptions define a one-dimensional problem that can be solved analytically or numerically, and the obtained results are normally presented in terms of *amplification function* $A(\omega)$, which represents, at each frequency, the ratio of the amplitude of the motion at the free surface to the amplitude of the motion at the bedrock or at an outcropping of the rock.

A typical result is shown in Figure 6.8, and it can be observed that the motion is amplified around given frequencies, corresponding to the natural frequencies of the soil deposit

$$f_n = \frac{(2n-1)\cdot V_s}{4H}, \tag{6.19}$$

H being the thickness of the soil deposit and n the *nth* vibration mode.

In particular, the fundament frequency (with $n=1$) is given by

$$f_1 = \frac{V_s}{4H}. \tag{6.20}$$

It can also be noticed that the motion is de-amplified in the high-frequency range due to the internal soil damping.

This basic result proves that one could expect that a structure with a natural period similar to the fundamental period of the soil would experience a more severe excitation and therefore would be more prone to damage.

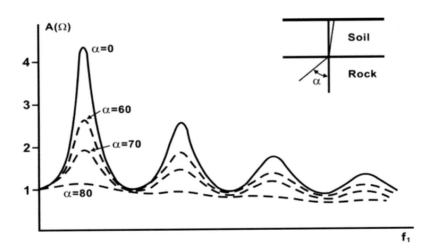

Figure 6.8 Soil amplification for shear waves horizontally polarized (SH) propagating at various angles from the vertical (Roesset, 2009)

These considerations form the bases of the so-called *microzonation studies*.

To evaluate the amplification ratio, if we assume that the internal damping of the soil at the fundamental frequency is equal to ξ_{soil}, the amplification from bedrock to the free surface can be estimated as

$$A(f_1) = \frac{2}{\pi \xi_{soil}}. \tag{6.21}$$

If we also take into account the loss of energy by radiation into the underlying rock by defining it as

$$\mu = \frac{\rho^s V_s^s}{\rho^r V_s^r}, \tag{6.22}$$

the ratio of the product of the mass density by the shear wave velocity for the soil to that of the rock, then it is possible to introduce an effective damping that accounts for both internal dissipation and radiation

$$\xi_{eff} = \xi_{soil} + \frac{2\mu}{\pi} \tag{6.23}$$

to be used into equation (6.21).

Note that the components of motion will always be amplified at the natural frequencies of the soil deposit, but the relevance of such an amplification will depend on the frequency content of the incoming waves, related to the magnitude and focal mechanism of the earthquake and the distance from the site.

As an example, large-magnitude earthquakes at long distance and long period of motion will have more significant effects at the top of a soft and deep soil deposit (therefore having a high natural period) than a smaller and closer earthquake having high-frequency components.

In addition, for more detailed studies, one should also consider the effect of the angle of incidence α of the incoming waves with respect to the vertical in the bedrock.

As shown in Figure 6.8 the amplification curves have similar shapes for different angles, but the predominance of the first peak has the tendency to reduce. Therefore, the simple one-dimensional model proves to be conservative.

A further aspect to be taken into consideration refers to the nonlinear behavior of soil.

In soil dynamics, it is common practice to highlight this aspect by plotting the secant value of the shear modulus as a function of the shear strain level (see Fig. 6.9).

Moving from this experimental evidence, Seed and Idriss suggested the use of an iterative linear procedure to define an equivalent linear system, and the effects of soil nonlinearity are apparent from Figure 6.10.

This figure shows the ratio of the response spectra of the motion at the free surface of the soil and the motion at the rock outcrop for two different levels of earthquakes, as expressed by the ground acceleration. It is then apparent that as the severity of the earthquake level increases, the spectrum tends to be broadened, the amplification reduces and there is a shift toward longer periods.

These studies in combination with recorded motions at different sites served to obtain seismic design coefficients, incorporating the effect of soil conditions, suggested in a number of codes.

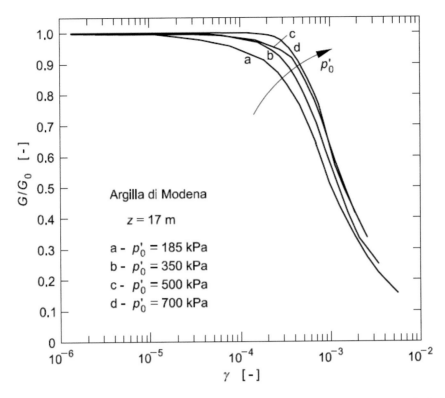

Figure 6.9 The decay of normalized shear modulus with the increase of the level of shear strain

Source: Lancellotta (2009b).

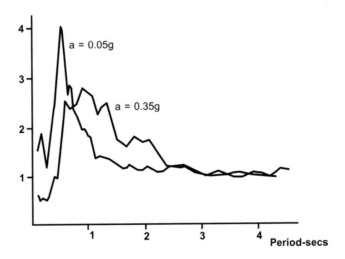

Figure 6.10 Ratio of response spectra for different levels of motion

Source: Roesset (2009).

These kinds of soil classification can help in routine practice, but it must be outlined that it rarely accounts for the natural frequency of the deposit and many other factors (magnitude of the earthquake, focal mechanism, distance from the fault, etc.).

Therefore, when dealing with built heritage, it is mandatory to rely on specific site condition studies by using natural records as motion inputs.

6.7 Summary of main points

When assessing the vulnerability of built heritage, it is of paramount importance to address the effects of local soil condition and properties on the response of the specific structure.

(a) This implies to consider a given number of design earthquakes at the bedrock or at an outcropping of rock.

These earthquakes, represented by *natural accelerograms*, should correspond to the magnitude, focal mechanism and distance for the site under consideration.

(b) The *amplification* at the free surface of the soil deposit can be computed as a function of the fundamental natural frequency of the soil deposit and the effective damping, which accounts for both the internal damping of the soil and the radiation into the underlying rock.

(c) The motion at the base of the foundation will be affected by the *kinematic interaction*, or wave scattering, and these effects will be significant for embedded foundations and foundations with large dimensions.

(d) Finally, the structure on soft soil has more degree of freedom and therefore different dynamic characteristics, and most of the vibrational energy is being dissipated into the supporting soil by radiation and by internal damping of the soil itself. Therefore, the effect of the *inertial interaction* is to increase the resonant period of the structure, with the benefit that the spectral acceleration to be used in design will be reduced; furthermore, the introduction into the analysis of the material damping of the half-space and the effect of energy radiation will result in an increase of the effective damping of the interacting system and a decrease of its response.

Chapter 7

Dynamic identification analyses

An approach to assess the real behavior of structures

7.1 Introduction

It was highlighted in Chapter 6 that to properly capture the behavior of a structure during seismic events, particularly of a historic tower, it is of paramount importance to take into account the interaction with the supporting soil and the effects related to soil conditions, that is, how soil properties may influence the input motion at the base of the tower.

The advent of powerful digital computers and the development of advanced numerical methods seems today allow to analyze any difficult interaction problems. However, the use of the so-called direct approach, which simulates the complete dynamic soil–structure interaction, requires a lot of expertise that is far beyond the current engineering knowledge. These difficulties motivated the introduction and the use of the so-called lumped parameters method (Sarrazin *et al.*, 1972; Kausel and Roesset, 1974; Kausel, 2010), rooted in the complete rigorous solution provided by Veletos and Wei (1971) and Luco and Westman (1971).

As we discussed in Section 6.4, in this approach the soil response is lumped into a series of frequency-dependent springs and dashpots, each of them related to the degree of freedom of the soil-foundation system (see Fig. 6.5).

Therefore, the procedure considers a "decoupled problem" to be solved in three steps (Gazetas, 1991; Kausel, 2010):

(a) the *kinematic interaction*, that analyzes the response of the foundation to the actual seismic motion defined in the free field,
(b) the formulation of the frequency dependent *dynamic impedances* for the foundation and
(c) the *inertial interaction*, that analyzes the structure supported on the impedances, defined in step 2 and subjected to the base motion obtained in step 1.

However, even considering this simplified approach, the designer will be faced with many critical aspects along his path, that have to be taken into considerations.

These include the complexity of the structure with all interventions operated along centuries, the unilateral behavior of masonry materials, the three-dimensional nature of the problem, the soil nonlinearity and its mechanical heterogeneity induced by the geostatic and applied stresses and the effectiveness of the soil contact along the vertical sides of the embedded foundation. Therefore, there is a need to complement this approach by means of experimental evidence to validate the assumptions used in analysis and design.

By no means is the best way to reach this goal to observe the structure as it behaves under dynamic excitations, and this is the essence of the so-called *dynamic identification analyses,* which allow the estimation of the natural frequencies of vibration, mode shapes and damping of the structure.

The experimental tests can be performed under known or unknown input. In the first case, the tests are performed using artificially produced excitation of harmonic type at variable frequency, usually by applying to the structure an eccentric mass rotating machine (vibro-dyne) or by producing a pulse load by means of a dropping weight. Excitations of harmonic type offer considerable advantages; that is, the different modes can be excited individually, and modal shapes and damping can then directly be identified and nonlinear behavior can be also explored, but these tests also involve several shortcomings. These include difficulties in setting up the testing equipment, the need to prevent the use of the structure during testing and very long execution times.

In contrast, dynamic tests with environmental excitation are performed through identification techniques that do not require a knowledge of the input, the results can be continuously recorded, and it is possible to perform an unlimited number of tests without impairing the integrity of the structure, an aspect of paramount importance when dealing with built heritage.

7.2 Basic aspects of the identification analysis

To show the capability of this analysis in a rather simple, introductory but effective way, let's explore the problem by considering in Figure 7.1 the monitoring system of the Ghirlandina Tower and consider at this first stage only the accelerometers at locations 3, 4 and 5.

Note that the location of the sensors on the structure should always be identified in advance by performing a sensitivity analysis by means of a simplified or a numerical model of the tower to detect the mode shapes and to avoid ambiguous estimates resulting from insufficiency of measurement points or by an incorrect distribution on these point on the structure.

Let now $y(t)$ be a time history of acceleration, $Y(f)$ its Fourier transform and $\bar{Y}(f)$ its complex conjugate. Then the *autospectrum* $G_{yy}(f)$ of each signal is defined as

$$G_{yy}(f) = Y(f)\bar{Y}(f), \qquad (7.1)$$

and the frequency response spectra in Figure 7.2 reveal peaks that can generally be assumed to represent either peaks in the excitation spectrum or normal modes of the structure (see Bendat and Piersol, 1993).

To distinguish peaks linked to the structural response from peaks in the excitation, we have to observe that points in a lightly damped normal mode of vibration will be in phase or 180 degrees out of phase, depending only on the shape of the normal mode. In particular, a distinction between bending and torsional modes requires phase measurements between two points on the opposite side of the structure but at the same elevation. In this respect, the autospectra of the signals do not provide information to identify the normal modes, and this is the reason why reference is made to the *cross-spectrum* $G_{yz}(f)$ function, defined as

$$G_{yz}(f) = Y(f)\bar{Z}(f). \qquad (7.2)$$

By referring to the illustrative example in Figure 7.3, the first peak indicates a bending mode, because measurements have a cross-spectrum phase equal to zero. On the contrary,

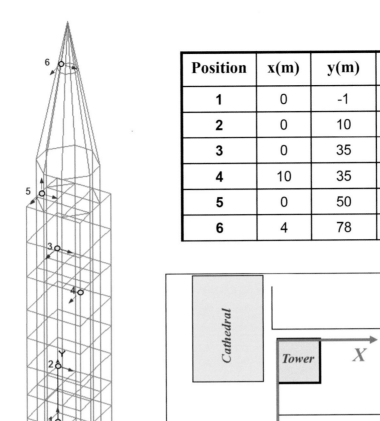

Position	x(m)	y(m)	z(m)
1	0	-1	0
2	0	10	0
3	0	35	0
4	10	35	0
5	0	50	0
6	4	78	4

Figure 7.1 Ghirlandina Tower: layout of measurement points

Source: Lancellotta and Sabia (2013, 2014).

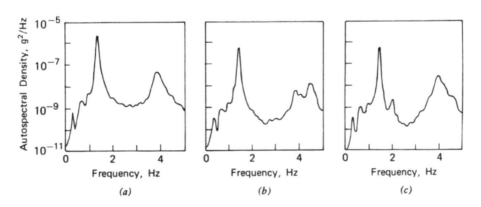

Figure 7.2 Autospectra of signals from three accelerometers: (a) and (b) located along the axis of the tower and (b) and (c) at the same level but opposite sides

Source: Adapted from Bendat and Piersol (1993).

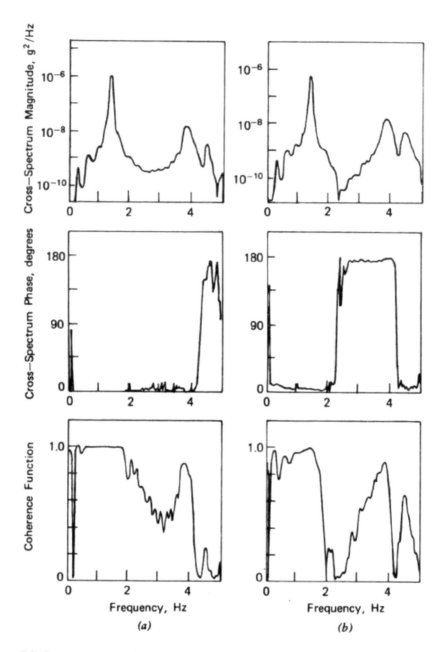

Figure 7.3 Cross-spectra and coherence functions between outputs of Figs 7.2 and 7.3:

    (a) point 3 *versus* point 5;

    (b) point 3 *versus* point 4

Source: Adapted from Bendat and Piersol (1993).

the second peak at a higher frequency suggests a torsional mode because the measurements along longitudinal separation are in phase, while points on the opposite sides are 180 degrees out of phase. The further peak would suggest a higher-order bending mode, but in this case the *coherence function*

$$\gamma^2(f) = \frac{\left|G_{yz}(f)\right|^2}{G_{yy}(f)G_{zz}(f)} \tag{7.3}$$

is as low as 0.25, that indicates that autospectra at this frequency include substantial extraneous noise.

Note that because this example is simply illustrative, the frequency values on the abscissa axis of Figures 7.2 and 7.3 have no real meaning, and those representative of the Ghirlandina Tower behavior will be described subsequently.

7.3 Identification analysis as applied to Ghirlandina Tower

The analysis of the seismic response of the Ghirlandina tower, measured during the earthquake of 3 October 2012 (with epicenter in Piacenza and Magnituto, M = 4.5) proves the capability of the previously introduced spectral analysis to detect the dynamic parameters of the structure.

As an example, Figure 7.4 shows the cross-spectra of acceleration time histories measured at points 1, 3 and 4 (see Fig. 7.2), and an inspection of this spectra reveals peaks associated with the structural response.

To extract from these measurements modal frequencies, shapes and damping parameters of the structure, a Stochastic Subspace Identification Method (Van Overschee and De Moor, 1996; Aoki *et al.*, 2007, 2008) was used, and examples of mode shapes, as detected on the plane (z, y) parallel to the cathedral nave and on its orthogonal plane, (x, y), are shown in Figure 7.5.

By referring to the modal shapes 1 and 2 in Figure 7.5, with associated frequencies of 0.74 Hz and 0.85 Hz, it can be observed that the first mode outlines the rotation at the tower basis due to soil deformability, whereas the second one claims for the presence of the arches connecting the tower and the cathedral.

In Figure 7.6 the first two modal shapes related to vibrations along the tower axis are also shown (modes 8 and 9), the related frequencies now being 4.51 Hz and 9.81 Hz, and here again the displacements at the tower base outline the contribution of soil deformability.

Therefore, a main conclusion may be reached that the soil–structure interaction cannot be neglected, in contrast to most identification analyses published in the literature that usually assume the structure with rigid constraint at its base.

The obtained results (mode shapes and related frequencies) are also of particular interest because they allow an estimate of the lower bounds of both the soil-foundation stiffness and the masonry modulus in a simple but effective manner.

Let first assume the tower as a rigid body resting on a deformable soil, then the frequency and the soil-foundation stiffness will be linked through the relation

$$f = \frac{1}{2\pi}\sqrt{\frac{K_{yy,lower}}{M}}, \tag{7.4}$$

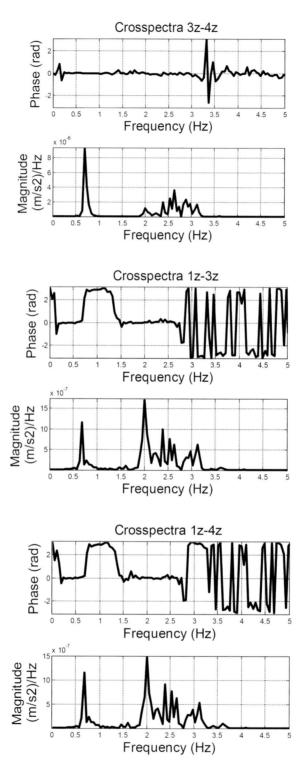

Figure 7.4 Ghirlandina Tower: cross-spectra functions among points 1, 3 and 4 of Figure 7.2

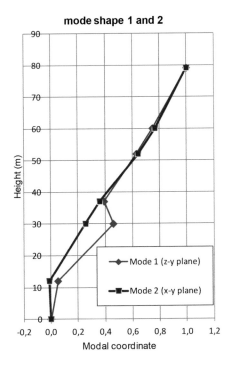

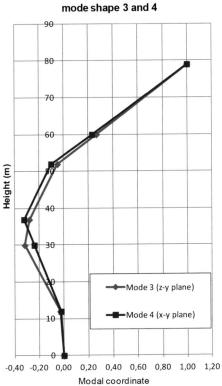

Figure 7.5 Bending modal shapes

Source: Lancellotta and Sabia (2013, 2014).

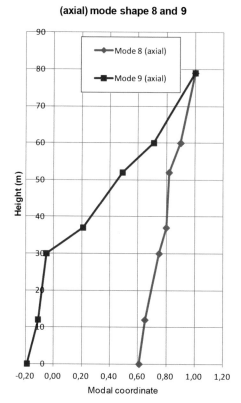

Figure 7.6 Axial modal shapes

Source: Lancellotta and Sabia (2013, 2014).

K_{yy} being the axial stiffness of the soil-foundation system (see Fig. 6.5) and M the mass of the tower. Because $M = 8525 \cdot 10^3 kg$ and the measured frequency is $f = 4.51\ Hz$ one gets

$$K_{yy,lower} = 6.84\ GN/m.$$

On the contrary, by assuming the soil as a rigid constraint while the tower is being represented with beam elements and lumped masses, a lower bound of the masonry modulus can be estimated, this being equal to

$$E_{m,lower} = 1.7\ GPa.$$

Obviously, soil and masonry deformability both contribute to the dynamic response of the structure, so an optimization analysis was performed by considering both bending and axial modes.

The obtained results prove that the first modes (in both cases of bending and longitudinal vibrations) are sensitive to soil stiffness, whereas the higher modes depend on masonry deformability, and we can optimize the response of the soil–structure system by using the masonry modulus E_m ranging from 3 to 4 GPa and the following values of the axial and rotational stiffness of the soil-foundation system:

$$K_{yy} = 7.8 \ GN/m$$

$$K_\alpha = 240 \ GN \cdot m/rad$$

Note that these values must be considered as representative of an elastic response, to be used with reference to serviceability limit states or in presence of low-intensity seismic events.

For this reason, the value of the rotational stiffness as deduced from identification analysis is in agreement with the theoretical value (Equation 6.18) that we can obtain when considering the shear modulus deduced from a shear wave velocity equal to $V_s = 125 \ m/s$ (see Fig. 5.3), the embedment of the foundation (5.65 m; that increases the value for shallow footing by a factor equal to 3.19) and the stress increment induced by the weight of the tower ($P = 85240 \ kN$). In the presence of strong motion, we need to properly take into account the decay of soil stiffness with shear strain, as it will be shown in the next section.

7.4 On the influence of soil nonlinearity

A key factor in the evaluation of the dynamic stiffness representing the soil-foundation response is the selection of an appropriate value of the shear modulus to be used in Equation (6.18) to define the impedance functions. In this respect, it can be expected to adopt a value of G consistent with the shear strain level as computed from a seismic site response analysis, as it will emerge from the experimental evidence discussed in the following. The same consideration applies to the strain-dependent internal damping, to be added to the radiation damping that accounts for the energy that radiates away from the foundation.

To study the role of soil–structure interaction in the long-term performance and to assess the seismic vulnerability of the tower, a dynamic monitoring program was started in August 2012. The acquisition system operates with a sampling frequency of 100 Hz and allows continuous monitoring of the dynamic response of the tower under ambient vibrations. In particular, the time histories of acceleration during recent earthquakes were also recorded. Specifically, three seismic events were recorded: 3 October 2012 (epicenter in Piacenza and magnitude M = 4.5), 25 January 2013 (epicenter in Garfagnana, M = 4.8) and 21 June 2013 (epicenter in Alpi Apuane, near Lucca, M = 5.2), and the comparison between the events of January and June 2013 is particularly significant to assess the influence of nonlinearity on soil response (see Fig. 7.7).

The same events have been recorded in the free field by the Modena station (MDN) of the Italian Accelerometric Networks, as shown in Figure 7.8.

By considering the relatively small distance between MDN station and the Ghirlandina Tower, if compared to the epicentral distances, this reference motion can be used to study the seismic response of the Ghirlandina site without the need to account for any attenuation effect.

The seismic ground response analysis was performed by using an *equivalent linear approach* (Idriss and Seed, 1968), as implemented in the numerical code EERA (Bardet et al., 2000). The reference shear wave velocity profile (Fig. 7.9) was estimated on the basis of the cross-hole test already discussed in Chapter 5.

Strictly related to this profile is Figure 7.10, which shows the change of normalized shear modulus and increment of damping ratio with shear strains, as measured on undisturbed

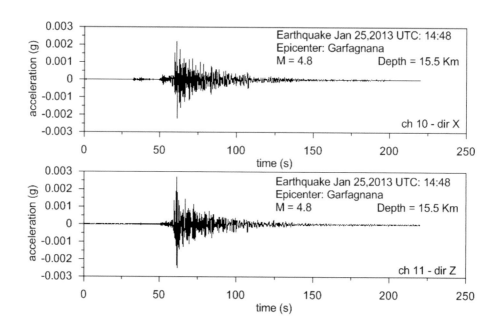

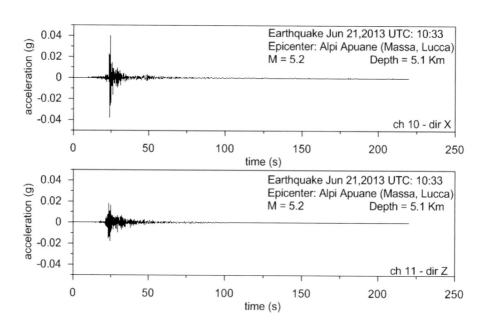

Figure 7.7 Acceleration records at the base of Ghirlandina Tower

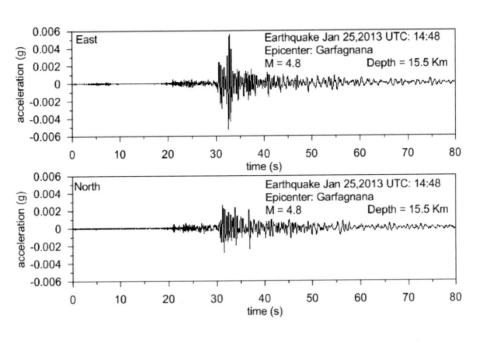

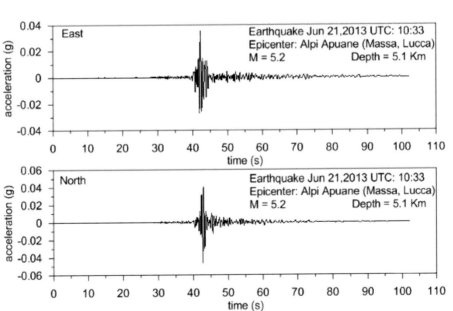

Figure 7.8 Acceleration records on the free field

samples with resonant column tests for the clayey layers. Note that for the sandy layers (letter *e* in Fig. 7.9), reference was made to the shear modulus reduction curves suggested by Seed and Idriss (1970). Also note that the sample obtained for the most superficial layer (letter *a* in Fig. 7.9) shows a lower linear threshold than other samples in agreement with its different grain-size distribution.

To perform the local response analyses, the free-field records have been deconvoluted to obtain the reference ground motions on the seismic bedrock, this latter assumed at a depth of 160 m, where the shear wave velocity is about 800 m s^{-1} (Fig. 7.9).

The obtained results are reported in Figure 7.11 for the three events recorded by the accelerometric station MDN. The profiles of maximum shear strains, if compared with

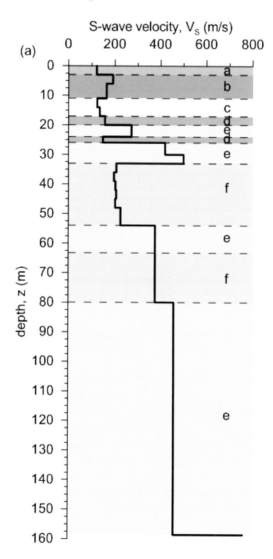

Figure 7.9 Soil profile used in local response analyses

Source: Cosentini *et al.* (2015).

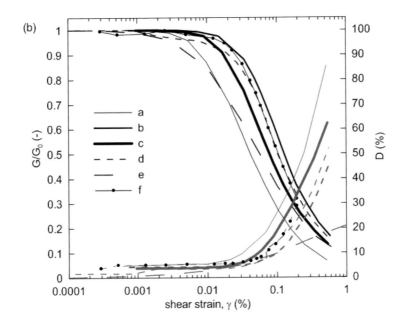

Figure 7.10 Dependence of shear modulus and damping ratio from strain level

Note: See Figure 7.9 for reference depths as labeled by small letters.

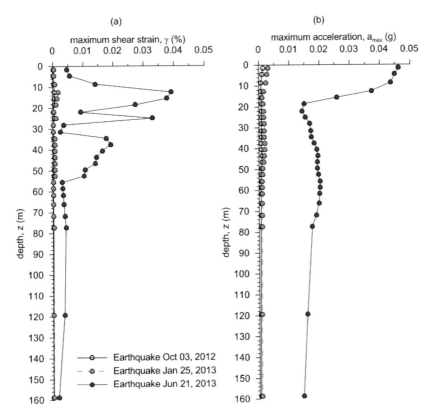

Figure 7.11 Soil response analyses

Source: Cosentini et al. (2015).

experimental reduction curves of the normalized shear modulus (G/G_0; Fig. 7.10), show that for the first two events, the response is associated with the behavior below the linear threshold strain. The third event, instead, caused larger strains, beyond the linear threshold, and in particular, the largest deformations are attained in the zone below the foundation, where a weak soil horizon is encountered. This layer is likely to have a major impact on the rocking response of the tower, and consistent with the shear strain level caused by the seismic motion of June 2013, an operative value of G/G_0 ranging from 0.75 to 0.8 is obtained within the zone of influence beneath the foundation.

7.5 Experimental validation of the procedure to select the "operational" soil stiffness

To validate the suggested procedure, an independent estimate of the reduction of foundation stiffness with increasing seismic action has been made on the basis of data collected through the permanent monitoring system of the tower.

The dynamic characteristics of the structure can be estimated in terms of the *transfer function* from the excitation at the base to the response measured on the structure at a given level. The *Frequency Response Function* (FRF) is therefore defined as follows:

$$H(\omega) = \frac{Y(\omega)}{X(\omega)} \quad \text{or} \quad H(\omega) = \frac{S_{XY}(\omega)}{S_{XX}(\omega)}, \tag{7.5}$$

where $Y(\omega)$ and $X(\omega)$ are the Fourier transforms of the output and of the input, respectively; $S_{XY}(\omega)$ is the cross-power spectrum between input and response; $S_{XX}(\omega)$ is the input power spectrum.

The FRF is a complex function which amplitude has a maximum at the resonant frequency, and the output is 90° out of phase with respect to the input.

Figure 7.12 compares the FRF, evaluated using the acceleration time history at the base as input, and the time history measured at 78 m as output for the three seismic events (October 2012, January 2013 and June 2013), and it is apparent there was a reduction in the first natural frequency of the tower (from 0.74 Hz to 0.69 Hz) moving from the first two events to the one of June 2013.

This difference is certainly associated with soil nonlinearity because significant structural nonlinearity is not expected for the masonry walls for such a small seismic excitation.

The observed difference in natural frequency can then be converted into an estimate of the reduction of the foundation stiffness by using the following arguments.

Let assume the tower to be represented by an equivalent single degree of freedom model, with a mass lumped at a height h over the base of the foundation and a structural stiffness equal to K_s.

If T_o is the fundamental period of the structure on a rigid base, it has already proved that the period of the structure–soil system increases when the flexibility of the soil is taken into account, and it is given by

$$T = T_o \sqrt{1 + \frac{K_s h^2}{K_\alpha}}, \tag{7.6}$$

where K_α is the rocking stiffness of the soil-foundation system.

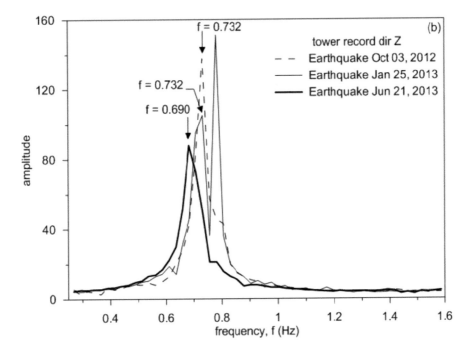

Figure 7.12 FRF for three seismic events

Source: Cosentini *et al.* (2015).

Provided that the value of T_o can be obtained by a numerical model calibrated on the structural identification process (in the present case $T_o \cong 1$ seconds), the ratio between the mobilized soil stiffness during two different seismic events can be obtained by using the inverse formula:

$$\frac{K_{\alpha 2}}{K_{\alpha 1}} = \frac{\left(T_1 / T_o\right)^2 - 1}{\left(T_2 / T_o\right)^2 - 1}. \tag{7.7}$$

From the difference in fundamental frequency observed in Figure 7.12, a stiffness ratio equal to 0.78 is obtained, consistent with the shear strain level derived from the seismic ground response analysis (Fig. 7.11).

At this stage we can summarize the main points as follows:

(a) Experimental identification analyses performed under ambient vibration excitation provide a sound validation of theoretical approaches suggested in the literature to estimate the dynamic stiffness of a soil-foundation system. However, considering the very low strain level involved in both identification analysis and shear wave propagation during cross-hole tests, the obtained values of the soil-foundation stiffness are appropriate only for low-intensity seismic motions. Further considerations need to be introduced when dealing with strong motions to account for nonlinear soil behavior.

(b) It is then suggested that a value of G consistent with the shear strain level computed from a seismic site response analysis should be used.

(c) Moreover, it is shown that the ground motion estimated from a local seismic response analysis represents a reasonable estimate of the foundation input motion to be used for structural analyses.

(d) Both conclusions (b) and (c) apply to the case of Ghirlandina Tower because they have been validated by means of a continuous monitoring system. This aspect deserves special attention because, as it was shown, monitoring increases the capability to detect the importance of nonlinear phenomena as well as possible damage that could occur in the long-term performance of the structure subjected to repeated seismic events.

7.6 A preliminary assessment of seismic vulnerability by limit analysis

A preliminary assessment of the tower seismic vulnerability may be performed by referring to limit analysis. This implies to consider a collapse mechanism, such as the one shown in Figure 7.13, the tower being loaded by horizontal forces that increase proportionally to a load multiplier λ (Fig. 7.14) until the mechanism takes place.

Figure 7.13 Collapse mechanism as overturning of the tower at the foundation level

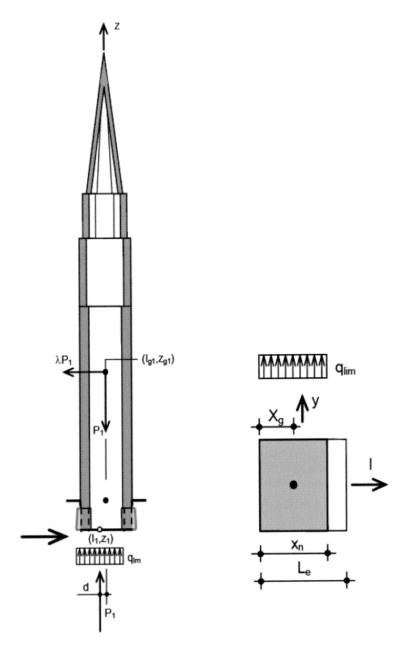

Figure 7.14 Collapse mechanism at foundation level

Source: Di Tommaso *et al.* (2013).

The mechanism shown in Figure 7.13 represents the overturning of the tower as a rigid body over a hinge at the foundation level, but obviously other mechanisms have to be considered at different elevations with hinges located where discontinuities in the cross sections of the tower are present. However, whereas local mechanisms can be prevented or contrasted by effective remedial measures, the global mechanism shown in Figure 7.13 is certainly the one of major concern.

When analyzing this mechanism, it must be outlined that, because of the finite value of the soil strength, as the multiplier is progressively increased, a bearing-capacity mechanism takes place well before an overturning mechanism is produced around a hinge located on the edge of the foundation.

In this condition, the contact normal stress on the foundation level is uniform and equal to the soil bearing capacity $q_{\lim}$ and a reduced width x_n must be considered to account for the load eccentricity (see Fig. 7.14).

If $P_1 = 85240\,kN$ is the weight of the tower, the equilibrium between P_1 and soil reaction allows the finding of the reduced width

$$x_n = \frac{P_1}{L_e \cdot q_{\lim}} = 7.50\,m, \tag{7.8}$$

with $L_e = 12.40\,m$ being the width of the foundation (see Fig. 5.2) and $q_{\lim} = 917\,kPa$.

When computing the bearing capacity, reference was made to undrained conditions, because earthquakes are short-term perturbations, and accordingly the analysis was carried out in terms of total stresses and undrained strength s_u, as it is usual in engineering practice. In particular, an operational value of $s_u = 125\,kPa$ was estimated by also considering aging effects that took place over at least 700 years following the construction of the tower (Pisanò et al., 2014; Mesri, 2014).

Once the reduced width x_n was defined, the maximum load eccentricity can be found equal to $d = 2.45\ m$ and by taking into account the existing leaning of the tower (1 degree) and the center of gravity height $z_{g1} = 30.61\,m$; the available net increase of the eccentricity in presence of seismic actions reduces to $1.92\ m$.

Consider now the tower to be represented by a single degree of freedom model consisting of a mass M lumped at a height h_{IN} over the base. As discussed in Chapter 6, this model can be thought of as representing the behavior of the tower in its first fundamental mode of vibration.

Accordingly, if x_i is the modal (or the assumed) horizontal displacement at the height h_i with a lumped mass m_i, then the modal mass associated with the first fundamental mode is given by

$$M = \frac{\left(\sum m_i \cdot x_i\right)^2}{\sum m_i \cdot x_i^2}, \tag{7.9}$$

and the height of centroid of the inertial forces is given by

$$h_{IN} = \frac{\left(\sum m_i \cdot x_i \cdot h_i\right)}{\sum m_i \cdot x_i}. \tag{7.10}$$

Therefore, the *seismic demand*, represented by the driving overturning moment, will be expressed by the quantity

$$D = S_a \cdot g \cdot M \cdot h_{IN} \qquad (7.11)$$

to be compared with the *capacity* of the system, represented by the stabilizing moment due to the weight of the tower

$$C = P \cdot d. \qquad (7.12)$$

For the present case, the modal mass is equal to $M = 5413 \cdot 10^3 \, kg$ and $h_{IN} = 39.94 \, m$; therefore, the maximum spectral acceleration will be equal to $S_a(g) = 0.075$, to be compared with the value expected at the site.

To forecast this expected value, a first step requires the selection of natural accelerograms representative of the seismic hazard at the site under consideration, which means considering magnitude, focal mechanism and distance from seismic sources.

Because these accelerograms are referred to bedrock or an outcropping of rock, the effects of the local soil condition and properties have been taken into account by performing a local seismic response analysis and the obtained response spectra, in terms of pseudo-acceleration are shown in Figure 7.15.

The effect of soil–structure interaction increases the resonant period of the structure, which can be evaluated by means of Equation (7.6).

Previously discussed experimental identification analyses performed under ambient vibration excitation provided a value of $T_1 = 1.35$ seconds, and this value was used to validate the soil-foundation stiffness.

However, as already discussed in Section 7.5, considering the very low strain level involved in the identification analysis, the obtained values are appropriate only for low-intensity

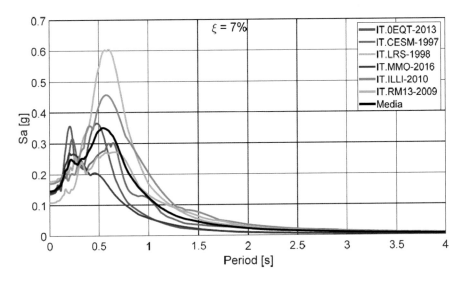

Figure 7.15 Normalized spectral acceleration for the expected design earthquake

seismic motions. Therefore, when dealing with strong motions, to account for the nonlinear soil behavior, a more representative value of the soil shear modulus G was selected, consistent with the shear strain level computed from the seismic site response.

To this aim, Figure 7.16 summarizes the shear strain level reached during propagation of seismic waves from the bedrock (fixed at depth of 160 m, where a shear wave velocity is about 800 m s^{-1}) to the surface, and it can be appreciated that, within the soft horizon that has a major impact on the rocking response of the tower, the strain level reaches values of at least 0.1%.

By considering the soil nonlinearity, usually expressed in terms of decay of the shear modulus with shear strain (see Fig. 7.17), a value of $G/G_o = 0.5$ can be expected to be representative for this strain level (note that in Fig. 7.17 the appropriate curve to be considered is that of the sample labeled as *clay 4* and *clay 5*).

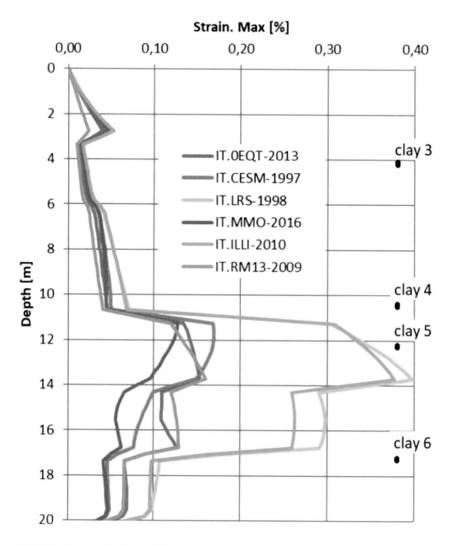

Figure 7.16 Strain level obtained from site response analysis

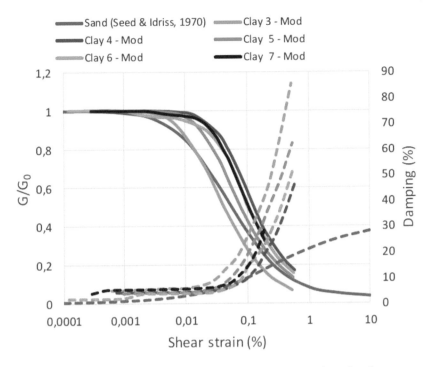

Figure 7.17 Decay of normalized shear modulus with increasing shear level

Finally, by using Equation (7.7) the following relationship can be established

$$(T_2 / T_o)^2 = 1 + (G_o / G_2) \cdot \left[(T_1 / T_o)^2 - 1 \right], \tag{7.13}$$

and this provides an increase in the fundamental period up to $T_2 = 1.63$ s.

The spectral acceleration corresponding to this period, as deduced from Figure 7.15, is on average equal to 0.036, and even considering the most adverse case it is lower than the value corresponding to the capacity of the system.

This preliminary assessment of the seismic vulnerability of the tower shows once again that the interaction of the structure with the soil has two important consequences:

It affects the seismic demand, as discussed in Chapter 6 because it increases the fundamental period of the structure and reduces the spectral acceleration;

in addition, it also affects the capacity of the system because this capacity is related to the soil strength and therefore to its bearing capacity.

References

Alfieri, S., Blasi, C., Carobbi, M. & Coisson, E. (2009) La struttura, dissesti e lesioni. In: *La torre ghirlandina un progetto per la conservazione.* Sossella Roma, vol. 1, pp. 146–163.

Aoki, T., Sabia, D. & Rivella, D. (2008) Influence of experimental data and FE model on updating results of a brick chimney. *Advances in Engineering Software*, 39, 327–335, ISSN: 0965-9978.

Aoki, T., Sabia, D., Rivella, D. & Komiyama, T. (2007) Structural characterization of a stone arch bridge by experimental tests and numerical model updating. *International Journal of Architectural Heritage*, 1, 227–250, ISSN: 1558-3058.

Baracchi, O. & Giovannini, C. (1988) *Il Duomo e la Torre di Modena*, Aedes Muratoriana Modena, Biblioteca-Nuova Serie n.105, pp. 19–28, 143–206, 253–263.

Baraldi, P. (2010). Indagine sui colori delle pitture murali della torre Ghirlandina. *La torre Ghirlandina: storia e restauro*, Sossella Ed. vol. 2, pp.172–177.

Bardet, J. P., Ichii, K. & Lin, C. H. (2000) *A Computer Program for Equivalent-Linear Earthquake Site Response Analyses of Layered Soil Deposit, User's Manual*, Los Angeles, University of Southern California.

Bendat, J.S. & Piersol, A.G. (1993) *Engineering Applications of Correlation and Spectral Analysis*. Wiley.

Bertacchini, E., Capra, A., Castagnetti, C., Rivola, R., Toschi, I. (2015) *La torre Ghirlandina, cronaca del restauro e studi recenti*, Sossella Ed. Roma (e-book) pp.21–23.

Binda, L., Gatti, G., Mangano, G., Poggi, C. & Sacchi, Mandriani G. (1992) The collapse of the civic tower of Pavia: A survey of the materials and structure. *Masonry Int.*, 20(6), 11–20.

Biot, M.A. (1932) *Vibrations of Buildings During Earthquake. Chapter 2 in PhD Thesis: Transient Oscillations in Elastic Systems.* Aeronautics Department, California Institute of Technology, Pasadena, CA.

Biot, M.A. (1941) A mechanical analyzer for the prediction of earthquake stresses. *Bull. Seismological Soc. Am.*, 31, 151–171.

Biot, M.A. (1942) Analytical and experimental methods in engineering seismology. *ASCE, Transactions*, 108, 365–408.

Biscontin, G. (2009). Il progetto per l'intervento: scelte e aspetti metodologici. *La Torre Ghirlandina: Un progetto per la conservazione*, Sossella Ed. Roma, vol. 1, pp. 236–241.

Brandi, C. (1977) *Teoria del restauro*. Einaudi, p. 154. (prima edizione, 1963, pubblicata nelle Edizioni di Storia e Letteratura).

Burland, J.B., Jamiolkowski, M. & Viggiani, C. (2003) The stabilization of the leaning tower of Pisa. *Soils and Foundations*, 43(5), 63–80.

Cadignani, R. & Valli, F. (2009) Il restauro degli anni Settanta. In: *La torre Ghirlandina: un progetto per la conservazione*. Sossella, Rome, vol. 1, pp. 88–93.

Calabresi, G. (2013) *The Role of Geotechnical Engineers in Saving Monuments and Historic Sites*. Kerisel Lecture, XVIII ICSMGE, Paris, pp. 71–83.

Calabresi, G. & D'Agostino, S. (1997) Monuments and historic sites: Intervention techniques. *Proceedings of International Symposium on Geotechnical Engineering for the Preservation of Monuments and Historic Sites*, ed. Viggiani, C., Balkema, pp. 409–425.

Cancelli, A. (1986) *Aspetti geotecnici della subsidenza. Ambiente: Protezione e Risanamento. Atti 2° Corso di Aggiornamento per tecnici di igiene ambientale.* Modena. Ed. Pitagora, Bologna.

Cancelli, A. & Pellegrini, M. (1984) *Problemi geologici e geotecnici connessi al territorio della città di Modena.* Atti 2° Congr. Naz. ASS.I.R.C.CO., Ferrara, pp. 53–64.

Carbonara, G. (2012). *Restauro architettonico: principi e metodo. M.E. Architectural Book and Review,* Roma, 207 pp.

Casari, E. (1984) *Osservazioni sulla planimetria del duomo di Modena: Lanfranco, i quadrati, le diagonali in Lanfranco e Wiligelmo, Il Duomo di Modena.* Panini Editions, Modena. pp. 223–226.

Castagnetti, C., Cosentini, R.M., Lancellotta, R. & Capra, A. (2017) Geodetic monitoring and geotechnical analyses of subsidence induced settlements of historic structures. *Struct Control Health Monitoring,* John Wiley & Sons, 24, 15, ISSN: 1545-2263, DOI:10.1002/stc.2030.

Cavani, F. (1903). *Pendenza, stabilità e movimento delle torri. La Garisenda di Bologna e la Ghirlandina di Modena.* Atti del Collegio degli Ingegneri e Architetti in Bologna.

Chopra, A.K. (1995) *Dynamic of Structures.* Prentice Hall, p. 729.

Clough, G.W. & Penzien, J. (1993) *Dynamic of Structures.* McGraw-Hill, New York, p. 634.

Colla, C. Pascale, G. (2010). Prove non distruttive e semidistruttive per la caratterizzazione delle murature della torre Ghirlandina di Modena. *La torre Ghirlandina: storia e restauro,* Sossella Ed. vol. 2 pp. 218–227.

Como, M. (1993) Plastic and visco-plastic stability of leaning towers. *International Conference Physic-Mathematic and Structural Engineering, in Memory of G. Krall,* Elba.

Como, M. (2010) *Statica delle costruzioni storiche in muratura.* Aracne, Roma, p. 902.

Cosentini, R.M., Foti, S., Lancellotta, R. & Sabia, D. (2015) Dynamic behaviour of shallow founded historic towers: Validation of simplified approaches for seismic analyses. *International Journal of Geotechnical Engineering,* 9(1), 13–29.

Cremaschi M., Gasperi G. (1989) L'alluvione alto medioevale di Mutina (Modena) in rapporto alle variazioni ambientali oloceniche. *Mem. Soc. Geol. It.,* 42, 179–190.

Croce, A., Burghignoli, A., Calabresi, G., Evangelista, A. & Viggiani, C. (1981) The tower of Pisa and the surrounding square: Recent observations. *Proc. X ICSMFE,* 3.

D'Agostino S. (2017) *Vulnerabilità e normative del patrimonio costruito storico, in Ingegneria e Beni culturali,* a cura di Salvatore D'Agostino, Il Mulino.

Desideri, A., Russo, G. & Viggiani, C. (1997) La stabilità di torri su terreno deformabile. *Rivista Italiana di Geotecnica,* 1(5).

Dieghi, C. (2009) Fonti e studi per la storia della ghirlandina. *La Torre Ghirlandina: un progetto per la conservazione,* 1, 48–65, Luca Sossella Editore, Roma

di Prisco, C., Nova, R. & Sibilia, A. (2002) *Analysis of Soil-Structure Interaction of Towers Under Cyclic Loading.* NUMOG 8, Roma, Pande & Pietruszczak, Balkema, pp. 637–642.

Di Tommaso, A., Lancellotta, R., Sabia, D., Costanzo, D., Focacci, F. & Romaro, F. (2013) Dynamic identification and seismic behaviour of the Ghirlandina Tower in Modena (Italy). *Proceedings 2nd International Symposium on Geotechnical Engineering for the Preservation of Monuments and Historic Sites,* Taylor & Francis, pp. 343–351.

Fazzini P., Gasperi G. (1996). Il sottosuolo della città di Modena. Accad. Naz. Sci. Lett. Arti di Modena, *Miscellanea Geologica,* 15, 41–54.

Ferri M., Ferraresi, Gelati A., G. Rossi, U. Tigges (2015). *Buche pontaie selettive per favorire i rondoni ed escludere i colombi dalla Ghirlandina. La torre Ghirlandina, cronaca del restauro e studi recenti,* Sossella Ed. Roma (e-book), pp. 54–58.

Gazetas, G. (1983) Analysis of machine foundation vibration: State of the art. *Soil Dynamics and Earthquake Engineering,* 2(1).

Gazetas, G. (1991) Foundation vibrations. In: Fang, H. F. (ed) *Foundation Engineering Handbook,* Van Nostrand Reinhold, New York, pp. 553–593.

Giandebiaggi, P., Zerbi, A. & Capra, A. (2009) Il rilevamento della torre Ghirlandina. *In La Torre Ghirlandina. Un progetto per la conservazione,* Ed. Sossella, Roma, vol. 1, pp. 78–87.

Hambly, E.C. (1985) Soil buckling and the leaning instability of tall structures. *The Structural Engineer*, 63(3), 77–85.

Hayashi, S.H., Tsuchida, H. & Kurata, E. (1971) *Average Response Spectra for Various Subsoil Conditions*. 3th joint meeting of US Japan panel on wind and seismic effects. UJNR, Tokyo.

Heyman, J. (1992) Leaning towers. *Meccanica*, 27, 153–159.

Holtz, R.D., Jamiolkowski, M. & Lancellotta, R. (1986) Lessons from oedometer tests on high quality samples. *JGED, ASCE*, 768–776.

Housner, G.W. (1954) Geotechnical problems of destructive earthquakes. *Géotechnique*, 4, 153–162.

Housner, G.W. (1959) Behaviour of structures during earthquakes. *ASCE, Journal Eng. Mech. Div.*, 85(EM4), 109–129.

Housner, G.W. (1970) *Design Spectrum. Chapter 5 in Earthquake Engineering*, ed. Wiegel, R.L., Prentice –Hall.

Idriss, I. M. & Seed, H. B. (1968). Seismic response of horizontal soil layers, Am. Soc. Civ. Eng. *J. Soil Mech. Found. Div.*, 94(4), 1003–1031.

Jamiolkowski, M., Ladd, C.C., Germaine, J.T. & Lancellotta, R. (1985) New developments in field and laboratory testing of soils. *Theme Lecture XI ICSMFE, San Francisco*, 1, 57–152.

Kanai K. (1957) Semi-empirical formula for the seismic characteristics of the ground. *Bull. Earthquake Research Institute*, 35.

Kàrmàn Th. von (1910) Untersuchungen uber Knickfestgkeit. *Mitteil. Forschungsarb.*, 81.

Kausel, E. (2010) Early history of soil-structure interaction. *Soil Dynamics and Earthquake Engineering*, Doi:10.1006/j.soildyn.2009.11.001.

Kausel, E. & Roesset, J.M. (1974) Soil-structure interaction problems for nuclear containment structures. electric power and civil engineers. *Proceedings ASCE Power Division Conference*, Boulder, CO.

Kramer, S.L. (1996) *Geotechnical Earthquake Engineering*. Prentice Hall, p. 653.

Labate, D. (2009) Archeology's contribution to understanding a monument. In: Cadignani, R. (ed) *The Ghirlandina Tower - Conservation Project*. L. Sossella Publisher, Roma, vol. 1, pp. 66–77.

Lancellotta, R. (1993) The stability of a rigid column with non linear restraint. *Géotechnique*, 33(2), 331–332.

Lancellotta, R. (2009a) Aspetti geotecnici nella salvaguardia della torre Ghirlandina. In: La Torre Ghirlandina (ed) *Un progetto per la conservazione*. Ed. Sossella, Roma, vol. 1, pp. 178–193.

Lancellotta, R. (2009b) *Geotechnical Engineering*. Taylor and Francis, London, p. 499.

Lancellotta, R. (2013) La torre Ghirlandina: una storia di interazione struttura-terreno. *XI Croce Lecture, Rivista Italiana di Geotecnica*, 2, 7–37.

Lancellotta, R., Flora, A. & Viggiani, C. (2018) *Geotechnics and Heritage. Historic Tower*. CRC Press, p. 261.

Lancellotta, R. & Sabia, D. (2013) *The Role of Monitoring and Identification Techniques on the Preservation of Historic Towers*. Keynote Lecture, 2nd International Symposium on Geotechnical Engineering for the Preservation of Monuments and Historic Sites, Napoli, pp. 57–74.

Lancellotta, R. & Sabia, D. (2014) Identification technique for soil-structure analysis of the Ghirlandina tower. *International Journal of Architectural Heritage*. DOI:10.1080/15583058.2013.793438.

Lomartire S. (2010) *L'apparato scultoreo e le fasi di costruzione della Ghirlandina. La torre Ghirlandina: storia e restauro*, Sossella Ed., vol. 2, pp. 60–142.

Lomartire, S. (2015) Il cantiere: lapicidi e scultori nella vicenda edilizia della torre. In *La torre Ghirlandina cronaca del restauro e studi recenti*, Luca Sossella editions (e-book) pp. 10–17.

Luco, J.E. & Westman, R.A. (1971) Dynamic response of circular footings. *ASCE, Journal of Engineering Mechanics Division*, 97(EM5), 1381–1395.

Lubritto C., Caroselli M., Lugli S., Marzaioli F., Nonni S., Marchetti Dori S., & Terrasi F. (2015) AMS radiocarbon dating of mortar: The case study of the medieval UNESCO site of Modena. *Beam interactions with materials and atoms*, Elsevier Ed., pp. 614–619.

Lugli, S. (2017) Mutina sepolta: inquadramento geologico dell'area urbana di Modena. In: *Mutina splendidissima, la città romana e la sua eredità*, De Luca editori d'arte, Roma, Novembre, pp. 16–19.

Lugli, S., Marchetti, Dori S., Fontana, D. & Panini, F. (2004) Composizione dei sedimenti sabbiosi nelle perforazioni lungo il tracciato ferroviario ad alta velocità: indicazioni preliminari sull'evoluzione sedimentaria della media pianura modenese. *Il Quaternario, Italian Journal of Quaternary Sciences*, 17, 379–389.

Lugli, S., Papazzoni, C.A., Rossetti, G. & Tintori, S. (2018) *Il paramento lapideo del duomo di Modena*, Collana Contributi per la storia materiale del Duomo di Modena.

Lugli, S. *et al.* (2009) Le pietre della torre Ghirlandina. In: La torre Ghirlandina (ed) *un progetto per la conservazione*, Sossella Editions, vol. 1, pp. 96–117.

Macchi, G. (1993) Monitoring medieval structures in Pavia. *Structural Engineering International*, 1, 9.

Marchi, M., Butterfield, R., Gottardi, G. & Lancellotta, R. (2011) Stability and strength analysis of leaning towers. *Géotechnique*, 61(12), 1069–1079.

Mesri, G. (2014). Discussion on Soil-foundation modelling in laterally loaded historical towers, by Pisanò et al., *Géotechnique*, 64(7), 587–588.

Morabito, Z., Tonon, M., Mazzari, M., Longega, G., Driussi, G., Biscontin, G. (2009) *La Torre Ghirlandina: Un progetto per la conservazione*, Sossella Ed. Roma, vol. 1, pp. 208–233.

Palazzi, G. (1988) *Analisi e interpretazione dei rilievi della facciata del Duomo di Modena in Appendice di Il Duomo di Modena Atlante grafico*, Carpi, pp. 155–157.

Papazzoni, C. A., Lugli, S., Pallotti, G. (2010). *Antiche tracce di vita riportate alla luce dal restauro. La torre Ghirlandina*: storia e restauro, Sossella Ed. vol. 2, pp. 64–67.

Peroni, A. (1984) *L'architetto Lanfranco e la struttura del duomo in Lanfranco e Wiligelmo, il Duomo di Modena (Quando le cattedrali erano bianche Mostre sul Duomo di Modena dopo i restauri)*, Panini Ed., Modena, pp. 143–163.

Piccinini, F. (2009) Note sul cantiere del Duomo e della Ghirlandina: Lanfranco, Wiligelmo, i Campionesi e il Comune medievale a Modena. *La Torre Ghirlandina: un progetto per la conservazione*, 1, 42–65, Luca Sossella Editore.

Piccinini, F. & Fiorini, T. (2015) La Ghirlandina: fasi di costruzione e varianti. In: *La torre Ghirlandina cronaca del restauro e studi recenti*. Luca Sossella Editions, (e-book), pp. 6–8.

Pisanò, F., Di Prisco, C.G. & Lancellotta, R. (2014) Soil-foundation modelling in laterally loaded historical towers. *Géotechnique*, 64(1), 1–15.

Puzrin, A.M., Alonso, E.E. & Pinyol, N.M. (2010) *Geomechanics of Failures*. Springer, p. 245.

Roesset, J.M. (2009) *Some Applications of Soil Dynamics*. The seventeenth Buchanan Lecture, Texas A&M University.

Roesset, J.M., Whitman, R.V. & Dobry, R. (1973) Modal analysis of structures for foundation interaction. *ASCE, Journal of Structural Division*, 99, ST3, 399–416.

Russo, P. (1985) L'abbassamento del suolo nella zona di Modena (1950–1982). *Tecnica Sanitaria*, XXIII, 293.

Sarrazin, M.A., Roesset, J.M. & Whitman, R.V. (1972) Dynamic soil-structure interaction. *ASCE, Journal of Structural Division*, ST7, 1525–1544.

Seed, H.B. & Idriss, I.M. (1970) *Soil Moduli and Damping Factors for Dynamic Response Analyses*. Report EERC 70–10, University of California, Berkeley, p. 40.

Seed, H.B., Ugas, C. & Lysmer, J. (1976) Site-dependent spectra for earthquake resistant design. *Bull. Seismol. Soc. Am.*, 66(1), 221–243.

Serchia L. (1985). *Studi e interventi sulla Ghirlandina. I restauri del Duomo di Modena 1875-1984 (Quando le cattedrali erano bianche, mostre sul Duomo di Modena dopo i restauri)*, Panini Ed., Modena, pp. 169–184.

Settis, S. (1985) "W pro V": la lettera rubata, in Wiligelmo e Lanfranco nell'Europa romanica. *Atti del Convegno*, 24–27 ottobre 1985, Modena 1989, 53–55.

Silvestri, E. (2017) *La cattedrale modenese preesistente all'attuale e le vicissitudini del cantiere lanfranchiano*. Atti e memorie serie XI volume XXXIX, Modena Aedes Muratoriana. pp. 23–42.

Shanley F.R. (1947) Inelastic column theory. *J. Aeronaut. Sci.*, 261.

Trifunac, M.D. (2006a) Biot response spectrum. *Soil Dynamic and Earthquake Engineering*, 26, 491–500.

Trifunac, M.D. (2006b) Brief history of computation of earthquake response spectrum. *Soil Dynamic and Earthquake Engineering*, 26, 501–508.

Van Overschee, P. & De Moor, B. (1996) *Subspace Identification for Linear Systems: Theory-Implementation-Applications*. Kluwer Academic Publishers, Dordrecht, Netherlands.

Veletsos, A.S. & Meek, J.W. (1974) Dynamic behaviour of building-foundations systems. *Earthquake Engineering and Structural Dynamics*, III, 121–138.

Veletos, A.S. & Wei, Y.T. (1971) Lateral and rocking vibration of footings. *ASCE, Journal of Soil Mechanics and Foundation Division*, 97, 1227–1248.

Vettese, A. (2009). *Come un telo per coprire i ponteggi può diventare protagonista. La Torre Ghirlandina: un progetto per la conservazione*, Sossella Ed. vol. 1, pp. 262–271.

Viggiani, C. (2017) *Geotechnics and Heritage*. 2nd Kerisel Lecture, ICSMGE, Seoul, pp. 119–140.

Appendix: introductory notes on the leaning instability of historic towers

As discussed in Chapter 1, the preservation of historic towers requires a deep understanding of their behavior and the reasons that allowed them to survive over the centuries. We outlined that the towers we observe today survived the initial period in which they could have been not so far from a *bearing-capacity collapse*, due to *lack of strength* of the soil. Delay or interruption of the building process allowed the foundation soil to improve its strength and the tower to be successfully finished. And due to uneven settlements, some of them appear today to survive at an alarming angle of inclination. This highlights the danger of a *leaning instability* (Heyman, 1992), which increases if there is a *lack of stiffness* of the soil (Hambly, 1985). In this respect, the leaning Tower of Pisa represents the most powerful example because the preservation of the tower was recognized as being a problem of leaning instability.

Moving from these considerations, this Appendix is intended to contribute to these aspects by introducing some notes on the problem of leaning instability of tall towers on weak soil.

A.I The inverted pendulum model

A convenient way to explore the leaning instability mechanism is to make reference to the inverted pendulum model (Figure A.1). The tower is represented by a rigid column of length h, connected at its base to a rotational spring.

The response of the soil-foundation system is lumped into this spring and strain-hardening plasticity models can provide a relationship that allows taking into account soil nonlinearity. Therefore, it can be expected that the resistant moment M of the soil-foundation system depends on the current tilt α of the tower and, as shown by many authors (Lancellotta, 1993; Como, 1993, 2010; Marchi *et al.*, 2011), we can express the nonlinear response of the spring as follows:

$$M = M_R(1 - e^{-(\alpha - \alpha_{\partial})/\gamma_R}),$$

(A.1)

where α_o represents any initial tilt, due to imperfections of the system or uneven settlements occurred during construction, and the quantity

$$\gamma_R = \frac{M_R}{K_\alpha^o}$$

(A.2)

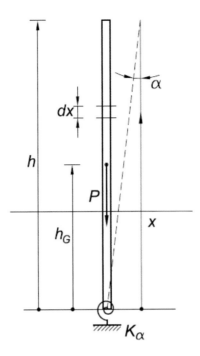

Figure A.1 The inverted pendulum model

depends on ultimate moment M_R (the asymptotic value attained for large α) and the initial rotational stiffness K_α^o of the spring.

We outline that the suggested relationship in Equation (A.1) implicitly depends on P, because $M_R = M_R(P)$, and provides a first important insight into the basic mechanism of the leaning instability.

The equilibrium condition in the deformed configuration requires

$$Ph_G \sin \alpha = M_R \left[1 - e^{-(\alpha - \alpha_o)/\gamma_R} \right],\tag{A.3}$$

and if the initial tilt is zero and the constraint has a linear behavior, by setting for small values of the current tilt,

$$\sin \alpha \cong \alpha$$
$$e^{-\alpha/\gamma_R} \cong 1 - \alpha/\gamma_R,$$

the equilibrium is assured until the critical load P_c is reached:

$$P_c = \frac{M_R/\gamma_R}{h_G} = \frac{K_\alpha}{h_G}.\tag{A.4}$$

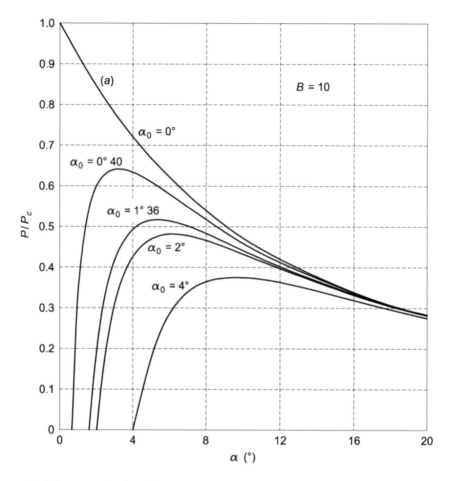

Figure A.2 Influence of soil nonlinearity on the post-peak behavior of a leaning tower
Source: Lancellotta (1993).

When the load is equal to P_c, the possibility of having neutral equilibrium requires a restraint with constant stiffness; on the contrary, in the presence of nonlinear restraint any increase of tilt produces a sudden collapse, as shown by the curve labeled a in Figure A.2.

A.2 The influence of an initial tilt and rotational creep

If there is an initial tilt α_o, the current tilt increases for any given load P and, due to the combined effect of nonlinear response and initial tilt, the collapse load is only a fraction of the critical load given by Equation (A.4).

Rotational creep, unlike vertical creep under constant vertical load, affects the load state of the foundation, and a tower, even if initially stable, can attain an instability condition with time.

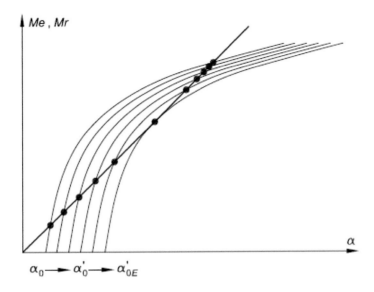

Me , Mr

α

α_0 ➝ α_0' ➝ α_{0E}'

Figure A.3 Time evolution of the stability condition

This can also be highlighted with reference to the plane in Figure A.3, where the external overturning moment $M_e = Ph_G \sin \alpha$ and the resistant moment M_r are plotted against the actual tilt (Marchi *et al.*, 2011). If the initial slope K_α^o of M_r is lower or equal to the slope (Ph_G) of the external moment load path M_e, then the M_r and M_e paths will never intersect and equilibrium is never possible. By contrast, if $K_\alpha^o > Ph_G$, for an initial α_0 a stable equilibrium condition occurs at the point on the M_r curve where $dM_r/d\alpha > dM_e/d\alpha$.

The maximum value of M_e is attained when the M_r curve is tangent to the M_e line, and this may eventually occur if we assume that the effect of delayed deformation (creep) is that of increasing the initial tilt α_0. Therefore, in this simplified interpretation, the effect of the creep is to merely translate the M_r curve along the α axis, while its shape is assumed not to be affected by creep deformation (see Marchi *et al.*, 2011).

A.3 Instability as a lack of stiffness

To reach quantitative conclusions, there are the following basic parameters to be assessed: the ultimate moment M_R and the initial stiffness. We can certainly rely on strength parameters when assessing the ultimate moment. On the contrary, although the values of the rotational stiffness can be defined by using a macro-element approach (see Lancellotta, 2013), it is rather difficult to validate the values so obtained by considering the three-dimensional nature of the problem, the mechanical heterogeneity induced by the applied stresses as well as the effectiveness of the soil contact along the vertical sides of the embedded foundation.

In this respect, a rather significant contribution comes from experimental identification analyses, due to the relation that exists between the results of these procedures and the dynamic nature of the problem we are dealing with, as it will appear subsequently.

As a first step, we have to recall that instability is a dynamic process, and for this reason we have to approach the problem from a dynamic point of view, because this is a necessary step to define the concept of stability. This approach will also allow us to explore the reason way we have previously analyzed the stability problem by neglecting its dynamic nature.

To answer this question, let consider again the inverted pendulum model (Fig. A.1) and analyze the dynamic equilibrium of the tower perturbed under the initial conditions

$$\alpha = \alpha_o \qquad \dot\alpha = \dot\alpha_o.$$

If ρ is the mass per unit length, by taking into account the inertial forces, the dynamic equilibrium writes

$$Ph_G \sin\alpha - K_\alpha \alpha - \int_o^h (x\ddot\alpha)\rho x dx = 0.$$

If, as usual, we consider small perturbations so that $\sin\alpha \cong \alpha$, we obtain

$$\ddot\alpha + \frac{3(K_\alpha - Ph_G)}{\rho h^3}\alpha = 0,$$

and we have to explore two cases:
  (a) If $P \cdot h_G < K_\alpha$, by introducing the quantity

$$\omega^2 = \frac{3(K_\alpha - Ph_G)}{\rho h^3} > 0,$$

the solution attains the following form

$$\alpha(t) = \alpha_o \cos\omega t + \frac{1}{\omega}\dot\alpha_o \sin\omega t.$$

In this case, the solution is bonded and, if the initial values are small, oscillations remain small for all times; that is, perturbations or changes in the system produce small oscillations and the quantity ω represents the circular frequency of the system.
  (b) On the contrary if $P \cdot h_G > K_\alpha$, the quantity

$$\eta^2 = \frac{3(Ph_G - K_\alpha)}{\rho h^3} > 0$$

and the solution

$$\alpha = \frac{1}{2}\left(\alpha_o + \frac{\dot\alpha_o}{\eta}\right)e^{\eta t} + \frac{1}{2}\left(\alpha_o - \frac{\dot\alpha_o}{\eta}\right)e^{-\eta t}$$

now prove that an instability condition has been reached because no matter how small the initial values are, the rotation becomes increasingly large.

Therefore, because the state of the system depends on the parameter $(K_\alpha - Ph_G)$, which represents the effective stiffness, this parameter varies the state branches at a critical value (a bifurcation or branching is possible) with a change of stability; that is, if the quantity $(K_\alpha - Ph_G)$ is positive the motion can be considered oscillatory with the frequency ω whereas it degenerates if the effective stiffness vanishes.

This means that once the critical load is attained, there is a *lack of stiffness* with respect to a possible deformed configuration, which superimposes to the trivial one, giving rise to a bifurcation of paths.

In addition to these remarks, the dynamic analysis suggests in a rather natural way the link between the frequency of oscillation and the stiffness of the system, so measurements of dynamic properties (frequencies, mode shape and damping) can provide the identification of the stiffness of the constraint, as discussed in Chapter 7.

A.4 Elastic-plastic regime

More insight into this problem can be reached by considering the two-degree model in Figure A.4. If it is assumed that the applied load is higher than the value corresponding to the yield load F_y of the two springs so that

$$P > 2F_y,$$

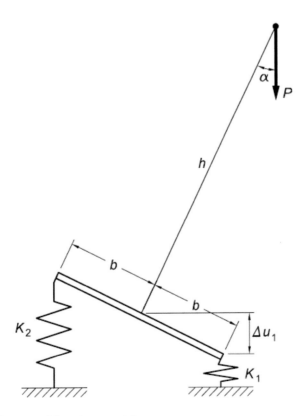

Figure A.4 Two degrees of freedom model

any small *perturbation at a constant load* will require the loading of spring 1, according to its elastic-plastic stiffness K_1, and the unloading of spring 2, as expressed by its elastic stiffness K_2.

The conditions to be fulfilled are

$$K_1 \cdot \Delta u_1 + K_2 \cdot \Delta u_2 = 0 \qquad\qquad (A.5)$$

$$\alpha = \frac{\Delta u_1 - \Delta u_2}{2b}. \qquad\qquad (A.6)$$

Stability against overturning also requires

$$P \cdot (h \cdot \sin \alpha) = K_1 \cdot \Delta u_1 \cdot b - K_2 \cdot \Delta u_2 \cdot b, \qquad\qquad (A.7)$$

and by combining the previous relations the following expression of the critical load in elastic-plastic regime is obtained, as first proved by von Kármán in 1910:

$$P_c = \frac{4b^2}{h} \frac{K_1 \cdot K_2}{K_1 + K_2}. \qquad\qquad (A.8)$$

An even more severe condition is attained if a load increment ΔP is applied so that the edge 2 does not move. In this case equilibrium requires

$$K_1 \cdot \Delta u_1 = \Delta P \qquad\qquad (A.9)$$

$$(P + \Delta P) \cdot (h \cdot \sin \alpha) = K_1 \cdot \Delta u_1 \cdot b, \qquad\qquad (A.10)$$

and the substitution of Equation (A.9) into Equation (A.10) gives the following expression of the critical load, first obtained by Shanley (1947):

$$P_c = \frac{2b^2 \cdot K_1}{h}. \qquad\qquad (A.11)$$

This expression gives a lower value of the critical load, as in this case it depends only on the plastic response K_1, whereas in the previous case there was the influence of both the elastic response (represented by K_2) and the plastic one.

Index

acceleration spectrum 98, 100, 127
Altare delle statuine (Michele da Firenze) 14
amplification (ground response analysis)
    104–105
Anselmo da Campione 7
ASCMo (historical archive) 45
auto-power spectrum 110, 122

Begarelli, Antonio (terracotta nativity) 14
Bellincini (chapel) 14
biological colonization 51
biological patina 53, 56
Biot, M.A. (response spectrum) 99
black crust 51–52, 55

Campionesi Masters 7
Canano (bishop) 14
capacity (bearing capacity of foundation) 126
Casari, E. 8
cathedral: geometry 8; history of construction
    7, 10
Cavani, Francesco 47
Cavedoni, Pietro 14
cleaning tests 59–60
coherence function 112–113
compression curve 84, 90
cone penetration testing (CPT) 86
Correggio 21, 29
Cristoforo da Lendinara, 14
critical load (leaning instability) 136
cross-hole test 82
cross-power spectrum 110, 122

damping: definition 95; material 102; radiation
    101–102; viscous 95, 102
damping ratio 97–98
di Duccio, Agostino 12
displacement spectrum 98
Dosso Dossi, 14
down-hole test 82
dynamic stiffness 102

dynamic stiffness of Ghirlandina soil-foundation
    system 117

earthquake (related to Modena site) 45, 47, 117,
    122
Enrico da Campione, 7
Eriberto (bishop) 5

fabbriceria 16
frequency (dimensionless parameter) 102
frequency of vibration (definition) 97, 105
frequency response function (FRF) 122

Geminianus, Saint 5
Ghirlandina: geometry 3, 17–19, 21, 80; self-
    weight 87, 126
ground response analysis 104
groundwater level 93

historical seismicity 45

identification analysis 109–110
impedance function 109
index properties (of Modena clay) 93
inertial interaction 104
integrity (concept of) 2
inverted pendulum model 135

Kanai 105
Karman (von) 99, 142
kinematic interaction 104

Lanfranco 5
laser technique (for crust removal) 69–70
leaning instability 135
leaning of Ghirlandina 3, 126
lithotypes (of Ghirlandina) 19, 49
local seismic response 104

massaro 16
Matilde of Canossa 7

Michele da Firenze, 14
Modena (city) 5
Modena clay 82, 85, 90, 93
Modena measures (perch, arm, foot) 8

normally consolidated soil 84

oedometer test 84
overconsolidated soil 84
overconsolidation ratio (OCR) 84

Paladino, Mimmo 49–50
Pala di San Sebastiano (Dosso Dossi) 14
Palazzo Ducale 39
parapet 13
Parma baptistery 13
period of vibration 97, 101
Piazza Grande 5–6
pollution (damage and deterioration) 44–45, 57
Porta dei Principi 11
Porta della Pescheria 13
Porta Regia 12
preconsolidation stress 84
pseudo-acceleration 98
pseudo-velocity 98

radiation damping 101–102
Relatio 6
resin (problems related to the use of) 43–44, 61
response spectrum 98
Rinaldi, Raffaele (il Menia) 21
rotational stiffness of foundations 102

Sandonnini, Tommaso 14
Secchia (room) 21, 23
seismic vulnerability 124
Shanley, F.R. 142
shear modulus 106–107
shear modulus reduction curves 107, 129
shear wave velocity 82–83
single degree of freedom system 95
soil amplification 104–105
soil-structure interaction 99
spire (of Ghirlandina tower) 25, 63
stiffness: of circular foundation 102; of
    Ghirlandina soil-foundation system 117
stress-history 83–84
subsidence 87, 92

terra verde 57
Torresani (room) 24
transept 12

undrained strength 126
UNESCO 2, 15

viscous damping 95
void ratio 83
vulnerability: intrinsic/induced 2; seismic 124

Wiligelmo 7, 10

yield stress 84

Zappolino (battle of) 21